Designing Mobile Service Systems

2nd Edition

Research in Design Series
Volume 2
ISSN 1569-7258

Previously published

Volume 1. R. Binnekamp, L.A. van Gunsteren and P.-P. van Loon, Open Design, a Stakeholder-oriented Approach in Architecture, Urban Planning, and Project Management

Designing Mobile Service Systems

2nd Edition

Els van de Kar

Alexander Verbraeck

in collaboration with

Wieke Bockstael-Blok

Mirjam Huis in ‘t Veld

Marijn Janssen

Carleen Maitland

Yan Wang

and

Students from Delft University of Technology,
Faculty of Technology, Policy and Management

IOS Press

ISBN 978-1-58603-919-6

Published and distributed by IOS Press under the imprint Delft University Press

Publisher and Distributor
IOS Press
Nieuwe Hemweg 6B
1013 BG Amsterdam
The Netherlands
Fax: +31 20 687 0019
email: info@iospress.nl
www.iospress.nl
www.dupress.nl

Distributor in the USA and Canada
IOS Press, Inc.
4502 Rachael Manor drive
Fairfax, VA 22032
U.S.A.
Fax: +1 703 323 3668
email: sales@iospress.com
www.iospress.com

PRINTED IN THE NETHERLANDS

Contents

Foreword

In the last few years it has become apparent that the introduction of mobile technology applications is not progressing in any way close to the forecasts and scenarios. A number of technological advances took place alongside the introduction of GPRS in Europe. For example colour screens, cameras and Multimedia Messaging Services (MMS) became available. Such features were first introduced in high-end smart phones, which operate with the Symbian OS that supports third party services. Java became more mature, which lead to the birth of a market for downloadable applications, in particular mobile games. Even streaming video to mobile phones became functional and it was generally believed that the new and better technology would promote and even drive both the emergence and the adoption of new mobile services.

3G is a generic term that describes the next generation of mobile communication and 3G refers in particular to high speed and multimedia data services. The introduction of 3G was intended to support a usage of high speed and multimedia data services and to bridge the gap between mobile phones and computing. The vision is – not surprisingly - the same as for the introduction of GPRS – higher speed networks, multimedia data services and the parallel use of multiple services should translate to more attractive mobile services, which will be adopted at increasing rates. As this did not happen with the GPRS there are now some elements of desperation and frustration in the new scenarios, and quite a few new services have been attributed to 3G but they have, in fact, already reached mature stages with the GPRS platform. Even services which rely on graphical browsing or multimedia messaging have approached basic availability for regular users. Information services, ticketing and different forms of entertainment are maturing services which can be used over a number of mobile technologies, including SMS (Short Message Service). Studies of the mobile Internet tend to neglect this because they do not consider SMS to be an Internet technology. The user is more concerned with the service than the technology and the basic form of the service may well be sufficient for the everyday needs of the user; more advanced technology may not be seen as value-adding by the user; (s)he may not even notice the difference as it is not considered relevant.

MMS is commonly considered to be an extension of SMS and should therefore represent a natural progression within the industry. What is obviously not well understood is that MMS is a much more complicated service to use and most users do not see enough added values over SMS to adopt MMS; what is needed is a value adding usage context. Mobile phones with wireless data capacity is another "inbred" design – as the use of the web is growing and the number of mobile phones is growing, the combination should be growing even faster; this is not the case, again we need a value adding usage context.

The basic challenge for GPRS, 3G and sets of new mobile services (which are being pushed into the market) is to understand how and why people adopt or do not adopt mobile services. In a series of studies of the Finnish consumer markets (carried out by IAMSR every year 2002-7) we have found that the relationship between technology adoption and the adoption of mobile services based on that technology is asynchronous, i.e. the adoption processes are different – the adoption processes can even be explained from different conceptual frameworks. A partial explanation can be found through the insights which are formulated in the so-called Braudel Rule. This rule was introduced in the Freedom Economy by Keen-Mackintosh (2001) and it states that "… freedom becomes value when it changes (actually "expands" in the original, French version; the difference is significant) the limits of the possible in the structure of everyday routines"; when this rule is applied to mobile services we could paraphrase it in the following way "… mobile services become mobile value services when they offer the possibility to expand the limits of the possible in the structure of everyday routines". This is a simple and effective way to formulate some guidelines for understanding what mobile services will create value and what mobile services will not make any difference. When applied to mobile service markets it may help us explain why some heavily promoted mobile services have failed, and why the SMS has been a success even if it was not advertised at all in the beginning – it was not even understood to be a mobile service.

Another partial explanation is, no doubt, that industry aims and consumer needs do not match; consumers are not part of the design process for the content/use context. Els van de Kar and Alexander Verbraeck have made a significant contribution to this part with the book they have authored: the theory and the methodology for design of service systems, which are enabled by new technology advances and innovations, is a crucial step for developing mobile value services. The design approach offers the means for gaining and systematising a better understanding of what mobile service customers want and need in the everyday users' contexts in order to "… expand the limits of the possible in the structure of everyday routines".

Prof Dr Christer Carlsson
IAMSR/Åbo Akademi University; Finland

Preface

This book is the second in the Research in Design series. Design is an effort that enjoys a growing attention in the academic world. At Delft University of Technology design is a recognized part of science. Like other technical universities, Delft is rooted in the engineering field. And in spite of questions like 'what is design', 'what is engineering' and 'what is science', which can be debated in long sessions, and differences that are hard to explain, it is possible to *feel* the differences. In this book we aim to accept the challenge of our Rector Magnificus' call (Fokkema, 2007) to contribute to the development of a design language; in this book for the service domain.

In general we observe that the engineering discipline is expanding into a field that embraces perspectives of more disciplines and actors, next to the engineer who is responsible for the artefact. The first volume in this Research in Design Series stresses the stakeholder oriented approach in the domain of architecture and urban planning (Binnekamp, van Gunsteren, & van Loon, 2006). The domain in this volume is *services*. This is a field in which the involvement of different stakeholders with different interests in the design process is particularly a critical success factor.

The design of service systems enabled by new technological opportunities is not easy. It is not always clear what users really want to do with e-services, development is costly, devices and available bandwidth change very fast, and business models for the different parties involved are not always clear. User interaction is also a problem: providing natural input is hard, and interfaces like screens might be small and only hold a limited amount of information. Furthermore, the fragmentation of services which requires to link services developed by different organizations remains a big problem. Therefore, there is a lot of interest in methods and best practices for the design of (mobile) services that are able to combine information from different data sources by ICT and target the information of different organisations at the individual wishes of the user. Mobile is put between brackets since this book can be used two-fold: as a book covering service systems in general and as a book focussing on mobile service systems. Our explanations, examples and illustrations are all from the mobile domain. The reason is that the challenges to overcome when designing mobile service systems are excellent examples for service systems design. Also, we believe that services will become ubiquitous and therefore the distinction between services and mobile services will become obsolete. The "anytime, anyplace" dimension promised in the beginning of the Internet Era is becoming reality with mobile services being part of our everyday life.

The theory in the book is based on the research on 'service systems engineering' as carried out between 2000 and 2005 in the Faculty of Technology, Policy and Management of Delft University of Technology. We would like to thank all staff members of this programme who contributed to the concept and notions of Service System Engineering. Especially we want to mention Gert-Jan de Vreede, Harry Bouwman, Rudi Westerveld, Marion Wiethoff, Erik Andriessen, Mariëlle den Hengst and René Wagenaar. We are very sorry that it is no longer possible to hand-over this book to René Wagenaar who always inspired us and recently passed away.

Nowadays, the service system domain is part of the Faculty research programs "Infrastructures" and "Multi-actor systems". In the Faculty's educational programs, we offer courses on 'Services Systems Engineering' to graduate students and many of the examples in this book have been based on assignments carried out by students in these courses. We are grateful to the lecturers and students who contributed to and used former versions of this book in their courses. A special thanks goes to Mark de Reuver who edited one of the former versions of this book.

Another inspiration for this book is the SmartAmlets© proposal that was submitted in October 2003 to the sixth framework programme of the European Union (Carlsson, 2003). Although the proposal was not granted, it was very useful as background for this book and we want to thank the programme coordinator Christer Carlsson from IAMSR, Åbo University in Finland. Furthermore, some of the material in this book has been based on the Ph.D. thesis of Van de Kar (2004). We thank all institutes and persons who participated in providing information for this book. Finally, we thank our co-authors: Wieke Bockstael-Blok, Mirjam Huis in 't Veld, Marijn Janssen, Carleen Maitland and Yan Wang.

To conclude we want to stress to our readers that we see the service design field as an emerging scientific area which is just at its beginning. We hope this book will be used in education, by practitioners, and by researchers.

A note on the second edition: some minor improvements have been made in the text and in the illustrations. The first and second edition are completely interchangeable.

Delft, June 2008
Els van de Kar and Alexander Verbraeck

1 Introduction

A key enabler for successful services is the move from a technology-centric world, where technology develops almost independently, to a user-centric world, where the development and use of technology originate from user needs. In this user-centric world companies with different backgrounds such as telecommunication, information technology, and media, have to cooperate to deliver services for these user needs.

1.1 What are Services and Service Systems?

Over the past decades, the nature of organisational work has changed. The manufacturing industry is no longer dominant. Today, the services industry, producing non-tangible goods, provides the majority of all jobs. In the OECD countries services in general account for 70% of total value added, and market services for 50% (OECD, 2007, p.48). In the USA more than 80% of the non-farming workforce was employed in the service-providing industry in the first quarter of 2007 (USA Bureau of Labor Statistics, http://www.bls.gov.). However, the amount of research aimed at the design, development and management of services is low. A key characteristic of the services industry is the high information content of its products, showing the potential for the application of information and communication technology (ICT). Where ICT was traditionally considered a driver for changes in organisational and inter-organisational processes, its role is now moving towards becoming an enabler for creating innovative service-related concepts. Innovation has been recognised as a key to growth, but the role of service-sector innovation has long been under-appreciated (OECD, 2007). It is not even clear how to measure service innovation, because service is also a very broad concept, see also (Tidd & Hull, 2003).

There are several definitions of a service, which highlight its main characteristics: "A market transaction, where the object is other than the transfer of ownership of a tangible commodity" (Judd, 1964), "Any activity or benefit that one party can offer to another that is essentially intangible and does not result in the ownership of anything. Its production may or not be tied to a physical product" (Kotler, 1997), and "A (series of) activities of more or less intangible nature that normally, but not necessarily, take place in interactions between the customer and service employees and/or physical resources or goods and/or systems of the service provider, which are provided as solutions to customer problems" (Grönroos, 2001). Existing definitions refer to specific characteristics of a service, e.g., the intangibility of the service, the interaction between producer and consumer when the service is created, the fact that a service cannot be stored after production to be delivered at a later time, and the fact that it is difficult to guarantee the quality of a service in advance.

Services are not only seen in an *inter-organisational* setting, such as business to consumer (b2c) services (e.g. a banking service using a mobile phone), business to business (b2b) services (e.g. a logistics service to organise the on time transport of goods between two places), and business to government (b2g) services (e.g. a service of the tax authorities to allow companies to file their tax statements electronically), but also as services *within organisations*. Examples of such services are services from financial departments to the rest of the organisation or services for internal teaching and training, also called business to employee services (b2e). A final category of potential successful services is in the consumer to consumer (c2c) domain; take SMS as an example of a well-known service in the c2c domain.

In this book, we view services from a systems engineering perspective, and therefore study *service systems* which include the user needs translated into performance criteria and operational processes, the information and communication technology to deliver the service, and the (inter)-organisational setting needed to develop and deliver the service. Designing effective service systems represents a challenge given the convergence in the service industry: the design object (the service system) represents an amalgamation of the service concept itself, the related organisational structure and the supporting information technology architecture. Focusing engineering efforts in an isolated fashion on one of these three elements is very ineffective, because the boundaries between the aspects of the service system are not clear and because they are reciprocally interdependent. Hence, there is a need for innovative engineering approaches, including design theories, design heuristics, modelling techniques and environments (laboratories) in which new organisational arrangements for services delivery can be tested and evaluated from a strategic, operational and technological perspective. In general, there is a need to integrate *soft knowledge* (e.g. the managerial perspective, trust issues, human factors) and *hard knowledge* (e.g. designing systems, identifying performance indicators, and developing state-of-the-art process simulations).

Although service jobs are predominant in western economies, the methods and tools for developing and assessing services are lagging behind in comparison with methods and tools to develop products and infrastructures. This can be seen in several application domains, where organisations have big problems in, for instance, defining new services based on high bandwidth mobile networks. In spite of the available bandwidth, and the availability of necessary hardware, the services that sensibly utilise this bandwidth are only appearing slowly. We see this, for instance, in the logistics and transport domain, where a shift is visible from purely distributing goods to offering door-to-door services that span several nodes of the supply chain. Transporters and goods are 'on the move', but the services to support real-time decision making in supply chains are almost unavailable. Services that really help the users with the right information at the right time in regular situations (e.g. during travel) or in exceptional situations (e.g. during a catastrophic event) are also hard to define, design, and deploy.

More recent studies focus more in the differences between processes involved in the provision of services on the one hand, and those involved in the manufacturing of products on the other. The differences turn out to be smaller than it appears. (Johnson, Menor, Roth, & Chase, 2000) argue that the distinction between service provisioning and manufacturing is blurred, while (Vermeulen & van der Aa, 2003) show that product development and service development have many aspects in common. The amount of research aimed at the design and management of services is rising (Amaoko-Gyampah & Salam, 2004; Grönroos, 2001; Menor, Tatikonda, & Sampson, 2002; Shulver, 2005; Stevens & Dimitriadis, 2005; Syson & Perks, 2004; Zeithaml, Bitner, & Gremler, 2006). This literature shows that the role of ICT is now changing towards an enabler for creating new service-related concepts. (Menor et al., 2002) argue that 'the role technology plays in New Service Design (NSD) is one critical area requiring further exploration'. One approach to do this is to consider services as systems that consist of different elements and, thus, to take the systems engineering perspective.

The above examples of service systems usually involve a number of organisations (or actors), and can therefore be characterised as inter-organisational systems. Inter-organisational systems are extremely hard to design, implement and manage, because there is not one 'governing' organisation that can command the others to comply with the same ideas and standards. A second factor that complicates the deployment of these services in practice is the inherent distributed character of inter-organisational services. Information and communication technology is needed to bridge the physical and cultural distances between the different organisations, both during the design and during the deployment phases. Standard ICT solutions to accomplish this are, however, still lacking, and the effectiveness of the solutions that exist is largely unknown, and therefore needs careful attention. Furthermore, the organisations involved are usually from different domains with their own habits and cultures. It takes time for these organisations to develop a common language. Strategic behaviour of the individual organisations might influence or even overrule the common interest of the network organisation.

1.2 Mobile Services Systems

In mobile service systems, all the components, such as handsets, content, applications, networks, user interfaces, and involved organisations, such as operators, content providers, service providers, have to work together to provide a mobile service to the users that adds value. If we combine the definitions on systems and services, and add the mobile component, the following definition for mobile service systems is derived:

> *A mobile service system is a group of components that work together for delivering a coherent set of activities of intangible nature that provide added value for a mobile user using a mobile network.*

This book looks especially at the challenges that organisations face when developing mobile, IT-based service systems that cross organisational boundaries and as a result cross time and space. Tafazolli et al. (2005) state in the strategic research agenda for eMobility in the European Union:

"Mobile and wireless communications, by the year 2020, is expected to play a central role in all aspects of European citizens' lives. The technology will substantially expand on the current concept of 'anywhere, anytime' to a new paradigm summarised in the following: 'Individual's quality of life improvement by making available an environment for instant provision and access to meaningful multi-sensory information and content'. This vision statement demands that the starting point of the design of future systems and services should be consideration of a person's basic needs and interests. This basic needs and interest involves comfort and welfare spanning one's personal, family, professional and private life. The technology should be all about improving the quality of life in terms of; wealth creation, improving education, improving job skills, enhancing health, security and safety and stipulating appropriate entertainment at the right time with appropriate content in a secure and reliable manner."

The realisation of innovative mobile service systems demands multidisciplinary research and development that crosses scientific and industrial boundaries. We aim to contribute to this by addressing the design issues for mobile service system from a socio-technical perspective.

Many of the problems regarding mobile, IT-based services can be solved by targeting the distributed nature and isolated deployment of mobile applications, by increasing the ease of use given the fact that screens and keyboards are small, and by providing business models that facilitate the linking of the applications and services of service providers when more providers are involved. One of the most promising ways to create viable services for mobile devices is to add extra intelligence to the services, both at the back-end and in the mobile device itself. The extra intelligence has to limit the cumbersome user interaction with the mobile device, speed up the handling of the transactions and ease the integration of different services that need to be accessed in order to complete more complex transactions. Intelligent services should be adaptive and intelligent ("smart") software modules, which are bundled from the perspective of the user and controlled by the context of the user. These software modules should be able to operate in a heavily distributed and multi-provider environment, and depend on mobile communication. With this type of services it should be easy for any person at any place, independent of preference, age, gender, education, background and nationality, to access services and complete transactions using a mobile device in an easy and efficient way.

In order to realise the rich mobile services landscape, the availability of intelligent software will be a critical success factor that can hide the complexity for the user, possess learning capabilities and so enable context aware personalised services.

Intelligent information management also implies a much higher level of user modelling than is currently the case, for the organisation, processing and presentation of data. Most current projects implement user *profiling* where the user's likely behaviour is inferred from a set of keywords. The challenge is to move to the next stage – user *modelling* – in which a much richer representation and processing capability will be required. Representation needs to be more than attribute – value pairs, for example hierarchies of interests, and user models that are themselves programs. It also needs to model "usual" as well as definite preferences – for example, a hotel booking system should know that I prefer a room that is single-bedded (normally, but this is not a strong preference), non-smoking (always) with breakfast included (but I like this to be an option to be decided according to my schedule).

Another aspect of the representation that needs to be modelled is time-varying preferences e.g. my browsing behaviour is different at weekends (leisure time) and during the week (business time). Finally the user model needs to be easily expandable for new services. Therefore it is necessary to develop an effective and scalable framework which can be embedded within any system to provide personalisation of system response, and understand ability and modifiability for the user. In particular this will include personalised hierarchical classification, so that content and services are presented in a manner convenient for the user.

At first sight, it may seem strange that it turns out to be so hard to create successful applications for high bandwidth mobile devices. The issue has been addressed in many publicly and privately funded projects, but, until now, with very limited success. One of the reasons is that mobile applications – often heavily distributed – are hard to develop. Another reason is that the business models are unclear: the organisation paying for the development is often not the one receiving the revenues from the use of the mobile application. Furthermore, the added value for the users appears to have been very limited until now: the small screen and small keyboard do not provide an easy user interface for more complex applications. A final reason is the isolated deployment of the applications, which force the user to access several applications sequentially to complete more complex transactions from the mobile device.

Breakthroughs are needed in three different fields, in order to make services intelligent. Firstly, the intelligence itself has to be designed, in such a way that it is flexible and can be easily deployed both on the client side and in the back office. Secondly, these adaptive, intelligent services differ so much from more traditional applications, that new design methods are needed that take the distribution of applications and providers into account. Finally, the bundling from the perspective of the user needs to take place. Only when the applications are aware of user needs and the user context, will it be possible to create added value for a particular user by taking away interaction steps that are 'standard' for that user.

1.3 Domains for developing mobile services

For this book, we chose example case studies from four domains that might benefit from mobile service technologies. Two domains are more dominated by personal requirements and two are determined by the requirements of organisations. We explicitly do not label it as consumer and business market since mobile services are mostly individually used, in private or work circumstances. Another distinction is the frequency of use: do people use the services continually or incidentally. We immediately admit that Table 1-1 is somewhat fuzzy since individuals might also be personally involved in crisis situations as a victim, workers need information, and so on.

Table 1-1 Example domains for mobile services used in this book

	Incidental use	Continuous use
Personal use	Travel support	Information & Entertainment
Organisation use	Crisis management	Workforce solutions

The first domain is *travel support*. Especially when 'on the road', outside of the well-known day-to-day context, users have a need for support. As fixed connections are seldom available to access all kinds of personal transactions and choices, travel support is a good domain to check the applicability and added value of mobile services. Take a business traveller who is at a meeting that takes longer than expected and has to fly to the next destination one day later. This person would like to book a hotel at the current location, re-schedule the flight, shift the taxi ride to the airport to the next morning (but then from the newly booked hotel and, dependent on the time of the new flight), cancel the hotel booking at the next location for one night, shift the taxi ride at the next location, and inform the partners at the next location that a delayed arrival is possible. With current mobile applications, it would take dozens of interactions and wading through long lists of flights, phone numbers and hotels to complete this simple 'change of delaying travel by one night'. Probably it would be easiest to call on human resources in the own organisation or at a travel agency, who would also need one to several hours to carry out this change. Given the fact that all the bookings and payment information are individually accessible in an electronic way and, in the near future, from any mobile device, it is absurd that such a transaction should take so much effort and so much time to complete.

The second domain is *mobile information and entertainment services* (MIES). These services have a high end-user context, and in Japan it has been shown that these services might well be the 'killer application', or the 'breakthrough service', especially for young people. Examples in this book show information on sports and cultural heritage, through which we can look at the differences between younger and older users in ease of accessing the services. Both b2c and c2c services can be considered here. The mobile information and entertainment service (MIES) domain is often quoted as a breakthrough application for mobile services, and the storyline could evolve as follows. A cyclist wants to do one of the famous classic cycle tours in Europe and gets travel descriptions dynamically on her mobile phone as she follows the route. On the road, she can get information about historical events that have taken place and when she passes a cultural heritage site which fits her interest profile she receives information and visiting options. Then the mobile phone can be used to pay for extra information or admittance fees. And since the classic cycle tours are organised tours she can always get logistic support from the tour organiser and immediate help, e.g. in case of an emergency.

The third domain is that of *mobile workforce solutions*. This is a business domain, and the users will be mobile workers. Mobile workers are considered to be people whose professional tasks can only be completed in a "mobile environment", which refers to the user's mobility and the need for technology that supports this mobility. Four fundamental aspects of mobile work are mobile workers, mobile tasks, mobile context, and mobile technology (Yuan & Zjeng, 2005). Examples of mobile workforces are salespeople in the field, travelling executives, people working in corporate yards and warehouses, and repair or installation employees who work at customers' sites. The nature of the services will be mainly b2b/b2e. One example is logistic chain management, where persons can use mobile services to speed up logistic processes, inform other partners in the chain about delays or problems, and jointly assess and solve these problems. Mobile workforce solutions can be used to simplify operations, carry out tasks in real time in order to reduce administrative routines and to give operators needed knowledge, provide information in the form needed, at the moment when needed and where needed, and ultimately reduce costs in each part of the logistic chain if it can be implemented in an effective and user supportive way.

The fourth domain from which examples are included in this book is *crisis management*. When we take emergency workers who are typical users of services in a crisis situation, we see, for instance, a clear need for flexible communication and for situational awareness. Because every crisis that emergency workers face is different, the services offered have to be quickly adaptable to the specific situation. What makes the design of end-user services for crises so difficult is that emergency workers can easily suffer from an information overload, and due to the nature of their work they have very limited space and time to interact with devices. In hazardous conditions, the potential use of IT devices is even more limited. When we add the complexity of overloaded networks during a crisis situation and the continuously changing picture of the crisis based on information that is slowly gathered by many different people and organisations, it is clear that the crisis management domain is one of the most difficult domains for service design.

1.4 Service Systems Framework

An ICT based mobile service system is part of a larger system. Systems are mostly associated with components, the relations between them and the purpose for which they are made. The purpose we are interested in in this book is to produce a service for end-users. We want to engineer, i.e. to design, make and maintain service systems. In this chapter we introduce a Service Systems Framework and the characteristics of mobile service systems. The three main aspects of this framework will be discussed in more detail in the next three chapters. Sections 1.6 and 1.7 will show two design examples of service systems designed by our M.Sc. students, to illustrate the service systems framework.

Service systems design is an extremely complex activity, and it encompasses several different disciplines. Therefore, we distinguish a number of different aspects that need attention when designing a service system. First, we need a *service concept*, which is the essence of the service which expresses how to create value for its users. Secondly, an appropriate *organisational network* should be in place that supports the (distributed) service production from an (inter)organisational point of view. The organisational network provides, for instance, the co-ordination between the resources, such as people and knowledge, in a (set of) service organisation(s). Thirdly, a *technical architecture* should be provided as well. For mobile service systems this includes of course information and communication technology. This technical architecture structures software, hardware, and netware that enable the delivery of a service. These three aspects, service concept, organisational network and technical architecture are of course very much interrelated, and are usually addressed concurrently.

For the total picture, see Figure 1-1. The picture also shows the interdependencies and trade-offs that need to be managed between the aspects. The technical and organisational network determines both the service concept that can or cannot be realised given the organisations that take part in the service system and the limitations of the technical architecture that is available to deliver the service. On the other hand, the organisational network and technical architecture also heavily determine the actual service delivery to the user.

The service system is positioned in an environment or application domain that is in a certain phase that can be labelled as trend. Furthermore, it might be dominated by other issues, the social, political and legal issues. This includes privacy issues, regulatory limitations, and ethical considerations. The economical aspects are also relevant. From the outside world developments on for instance the stock-market and the labour market influence the service system. Inside the service system financial aspects have to be taken into account. (Faber et al., 2003) include the financial arrangements as a separate aspect of business models for services in their STOF framework. In our framework, we do not separate the financial aspect as another element, and the pricing model for the customer is part of the service concept; revenue models between partners are part of the organisational network, and costs are part of all three aspects.

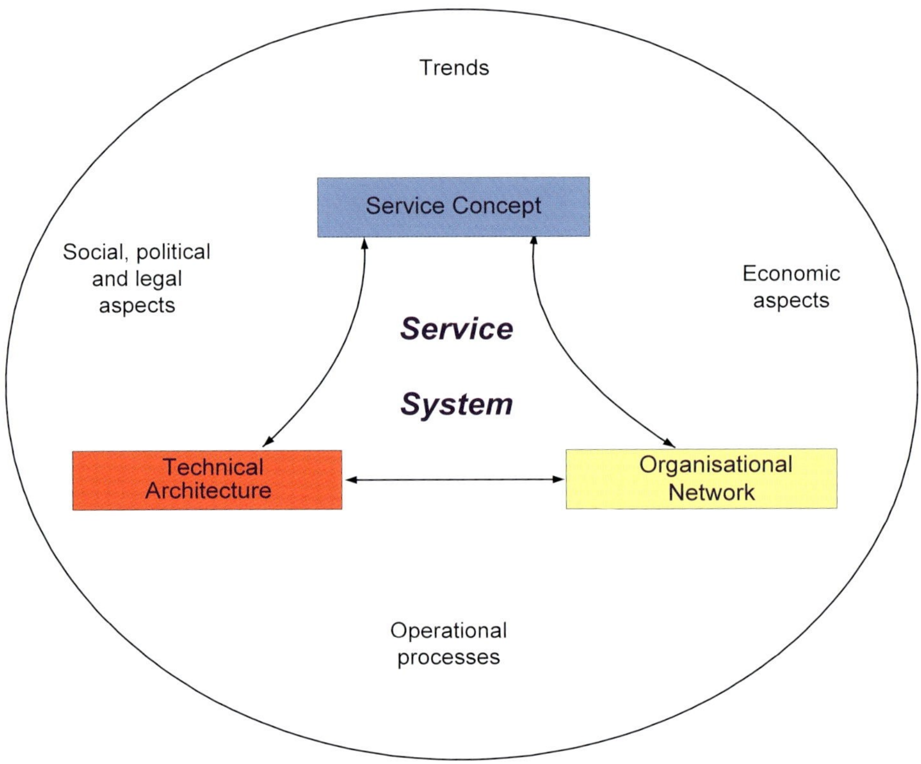

Figure 1-1. Distinguished aspects of a service system

In the service innovation model (Hertog, Broersma, & Ark, 2003) four dimensions for service innovation are distinguished: the technology dimension and three non-technology dimensions: the service concept, the client interface and the service delivery system. In our framework the first two elements are integrated into the service concept. The delivery system of the service is related to operational processes behind services. It depends on the system boundaries which aspects have to be taken into account. Designers choose the system boundaries based on the relevant issues in the specific application domain of their service system, e.g. a health care service or a transportation service.

Mobile service systems are closely related to distributed information systems, and as such the service systems engineering field can learn much from the distributed or inter-organisational information systems field when designing services systems, as we will see in chapter 5. There are, however, also several differences. Where information systems are usually aimed at realising the (support for) operational processes, the service system starts with a service concept that is offered to a set of (potential) customers. The service systems designer focuses on the added value that can be delivered through carefully designed interfaces for the users of the service. Because services are aimed at delivering added value through their interfaces, it is possible to link services and to have parallel services that are delivering the same or similar content through different channels. Think of the banking service of providing bank account information: the same information from the same underlying banking information system can be delivered on paper, by phone, through on-line banking or Internet access, and through an Automated Teller Machine. This example also shows an interesting link between the two fields; that service systems are usually built on top of sets of existing information systems that are already in place in organisations involved in the service delivery.

The challenge of designing mobile information service systems is a combination of the design of service systems and the design of information systems in an inter-organisational setting. During this design process choices have to be made. These choices concern the trade-offs between the aspects of the service system as shown in Figure 1-1. An example of such a trade-off is the choice between the position accuracy of a location-based service demanded by the user, and the position accuracy that can be provided by the available reliable technology (accuracy versus reliability). These trade-offs have to be made between two aspects but sometimes also between three aspects, within one aspect or between the aspects and the environment. It might be more efficient to start testing a best effort solution than to continue searching for an optimal solution. In the next three chapters we will further elaborate on the three main aspects of the service framework: service concept, technical architecture and organisational network. To prepare for this more in-depth discussion, we will discuss the typical characteristics of mobile service systems in the next section.

1.5 Mobile services systems characteristics

A service system is aimed at providing added value for its users. This added value is realised when users *can use* the service system, and really *do use* the service system. This usage of the service is part of the service concept, enabled by a good technical architecture, and offered by a good organisational network. The design process of the mobile service system can start at any of these three aspects. The service system might be developed based on an idea for a new service concept that adds value to users. The service system might be developed because a new technology has arrived from the lab. Or an organisational network has changed and therefore the service system had to be re-designed. Depending on the 'triggering aspect' the requirements for the service system will change.

The usage of mobile services implies costs and benefits. This holds for companies as well as for consumers. For companies that want to incorporate mobile solutions in their business processes the value can be made explicit by the identification of fundamental objectives (Nah, Siau, & Sheng, 2005). Nah et al. consider the following as the most important fundamental objectives: efficiency, effectiveness, customer satisfaction, security, cost and employee acceptance. Deighton (2003) indicates in a Gartner study the costs to achieve the potential benefits of mobile workforce solutions. Wang et al. (2005) summarised these benefits and costs for workforce solutions and we further extended it for mobile services in general; see Table 1-2.

Table 1-2. Benefits and Costs of Mobile Services

	Benefits	Costs
Tangible	▪ Save time or effort ▪ Business process efficiency/productivity ▪ Information accuracy- reduced correction costs ▪ Asset tracking ▪ Eliminating fixed infrastructure	▪ Capital hardware, device, servers, software ▪ Operational cost: especially airtime ▪ Software licenses ▪ Support and management ▪ Integration e.g. with legacy systems

Intangible	▪ Efficiency: convert dead time into useful time ▪ Happiness: always connected and entertained ▪ Employee quality of life e.g. flexibility, in touch with family ▪ Customer satisfaction ▪ Brand image & value ▪ Flexibility & agility ▪ Accessibility, presence and responsiveness	▪ "Always on" stress ▪ Destruction of communities of practice ▪ Complying with legislation e.g. working hours ▪ General attitudes and resistance ▪ Nuisance in public places by ringing phones, loud talking people ▪ Privacy ▪ Last-minute attitude to deadlines

We will illustrate the design of a mobile service system with two design examples, 'Guido the Guide' and 'Relief workers'. For both design examples we will describe the three main aspects of the framework of Figure 1-1, illustrating how these aspects can be described and how they relate to each other. The 'Guido the Guide' design example is triggered by a new service concept and the 'Relief workers' design example is triggered by opportunities to improve an existing service system that is already being used.

1.6 Design example: Guido the Guide

This design example concerns a service system that is developed for the travelling domain. The service system assists tourists who are in an unfamiliar city, and have some spare time. It gives suggestions to these tourists on how they can spend this spare time, taking personal preferences and the context of the user into account. Personalisation and localisation are thus important issues in the design example.

1.6.1 Service Concept

Imagine that you are a tourist in London and that you just visited Tower Bridge. In the evening you have to visit your uncle in Notting Hill. Now, you are looking for an activity for the afternoon. You take your mobile device and ask Guido the Guide to propose an interesting programme for the afternoon according to your known preferences. The service knows where you are now and where you have to be tonight and at what time. With only a few clicks the service presents a list of possible routes for the afternoon. Each programme consists of an interesting set of activities that meet your personal interests. You select the programme you like best. Then, the mobile device tells you where to walk or tells you which public transport you need. In the latter case it will lead you to the nearest station. While following the route, the service gives you some facts about the points of interest you pass on your route.

When the Guido the Guide service generates the possible programs it takes several factors into account. Firstly, it knows where you are. By looking up your schedule it knows where you have to be in the evening. Of course, if you do not have any appointments that day, it would be possible to specify a location and end time yourself. The service will plan a route that has a starting point, your current location, and will end at the location where you have to be tonight. Of course it will take the time constraints into account as well.

As mentioned before, the service knows your personal interests. These are not only used to select points of interest for your along the way, but also in planning the routes. Imagine that it is 11:00 hours when you ask your mobile device to devise a programme for you. The service knows that you usually have lunch between 12:00 and 13:00. It will take this into account and plan a stop at a restaurant at your usual lunchtime. If you like fast food more than an Italian lunch, it will automatically search for a snack bar along the route. If there is an attraction that exactly fits your personal preferences, but is far away, then you do not have time to lunch at a restaurant. It will also give this route, but puts it lower in the ranking. Many other aspects can be taken in account, such as the weather forecast. If it knows you like theatres, it will search for an interesting play, see whether tickets are still available, and if you confirm you want to go there, the service will book a ticket for you.

User process

In order to show how the service will work, a sketch of the user interfaces will be given. This is done by the use of a storyboard.

It all starts with a person in a new city. And some spare time... What to do? Fortunately, Guido the Guide can help:

With several clicks the user starts the service:

Guido presents several routes to the user. The user chooses the route from the list he/she likes best. The route starts by presenting a map, and telling the user where to go...

When the user walks by some interesting attractions (according to the personal profile) he/she immediately gets relevant information. As a result, the user follows a complete route, finishing at the right time and location.

1.6.2 Organisational network

Many information sources and sub-services are needed to fulfil the requirements for this service system. Where does all the information come from? Which sub-services are needed? Who is responsible for which part of the information? Many parties have to collaborate to make this service a success. How should this be arranged? And how can this result in a viable service? The following actors can be distinguished who should make arrangements and distribute costs to deliver this service:

Guido the Guide service organisation. This is the actor that will set up and manage the service. This firm will be responsible for the functioning of the service. The "Guido the Guide" users will be clients of this actor, and therefore pay this actor. This actor has to make sure that the service works properly, and to deliver the correct information to the user.

GIS (Geographic Information System) provider. The information about the location of the user will be retrieved using either GPS positioning or the positioning capabilities of the telecommunications network. Both will produce x-y coordinates. These coordinates need to be translated into an address, postal code or a map. To do this, a GIS provider will be needed.

Weather information provider. The service will make use of the local weather forecast. At least the service needs to have an answer to the question whether it will rain or not. Therefore a weather information provider will be needed.

Public Transport information provider. The service may advise making use of public transport. Therefore it needs to know where stations are, what the lines are, how long transportation will take, and so on. There are many service providers who can give this kind of information. For every city in which the service is available a Public Transport information provider has to be contracted.

Tourist information providers. To make this service work, information about attractions, restaurants, cinemas, and so on in a city is needed. This might be one service that is already available in the city, or a collection of services from different providers with whom separate contracts might be made.

Booking service provider. The service needs to have access to a system to book tickets for the cinemas and theatres. Therefore, a party will be needed that provides such a system, for example via the Internet.

Mobile network provider. The whole service will be presented on a mobile device to the user. This mobile device has to make contact with the server. Therefore it makes use of a GPRS or UMTS connection that is provided by a mobile network provider. To make sure that the user can use the service in any country, roaming agreements will be needed between different network providers. In a case where the positioning takes place using the positioning capabilities of the UMTS network, this actor will also provide the x-y coordinates of the user.

1.6.3 Technical architecture

This section briefly describes the technical architecture of the service system, see Figure 1-2. The user accesses the service with his mobile device. The user is in a city and the mobile device sets up a connection to a local telecom operator. When the user is abroad, a roaming agreement is needed between this local telecom operator and the user's home operator. Via the UMTS or GPRS connection, the user is able to access the Internet. By this he/she can log on to the main server. Of course, this all happens by just selecting this service from a favourites list.

The main server, that hosts the service, is also connected to the Internet. By the use of the Internet the service makes connection to all sub-services needed to provide the user with the route information. The main server also consists of a database with the personal profiles of the known users.

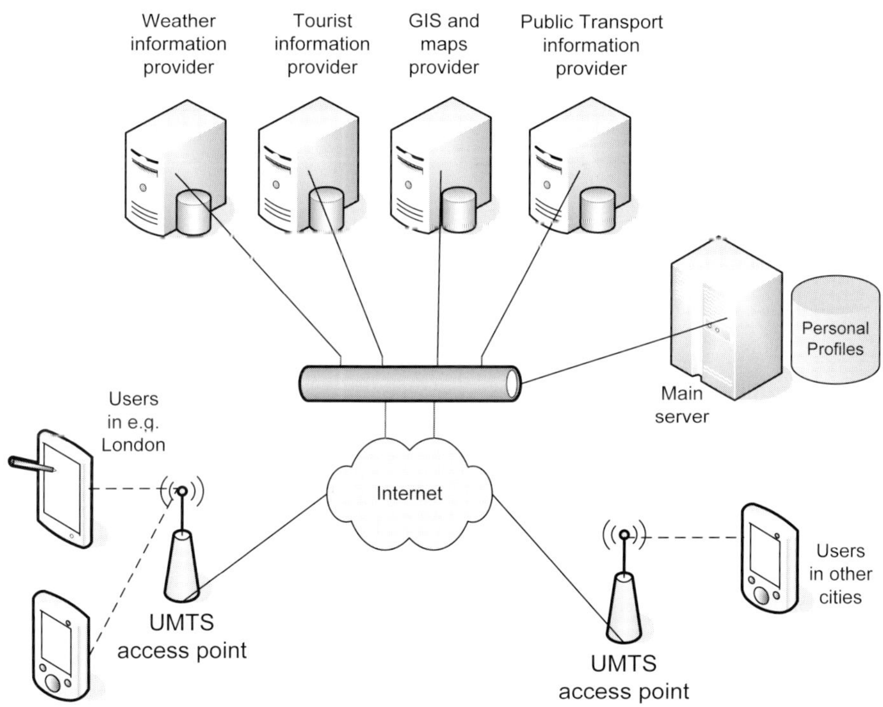

Figure 1-2. Technical architecture of the "Guido the Guide" service

1.6.4 Result

This design example illustrated how a service system can be developed by starting with the context of a specific user. From this point on, first the concept of the mobile service is designed and next the other elements of the service system are determined.

1.7 Design example: Port Emergency Response System

This is a design example in the crisis management domain. The context of the service system that is described consists of the activities of relief workers in the Port of Rotterdam. In a large port, there is always a risk that a major crisis can take place, for example the breakout of a fire in a location where fuel is stored. Once relief workers are present at the scene of such an emergency, they have to know what to do and they need to communicate and learn about the location of the disaster. The service systems that are currently used to do so have significant shortcomings. In this design example, a new service system is described that aims to solve these shortcomings, by providing real-time information to relief workers from different existing service systems and data sources about the emergency situation and on how to respond to it.

This design example illustrates how a service system can be designed when starting with the shortcomings of an existing service system.

1.7.1 Organisational network

The responsibility for crisis response in the Port of Rotterdam is located in one office, called the Rotterdam Port Authority (RPA). One of the mission critical tasks of the RPA is the mitigation of crises and disasters in collaboration with other teams.

Once a crisis exceeds a certain magnitude, and cannot be considered a routine accident that can be solved independently by one or several different service organisations, actors from these organisations join to form a Commando Place Incident team (CoPI team). The CoPI team should be read as a local crisis response team. The core tasks of a CoPI team are (Appelman & Van Driel, 2005):

- To coordinate the deployment of crisis mitigation units in the area of the incident
- To guide and direct the operational execution of all involved teams
- To take measures aimed at the prevention of more casualties

- To provide information about the repression of the crisis to a higher-level Regional Operational Team (RegOT)
- To request potential support from RegOT

In case of an emergency in the port of Rotterdam, the following four important actors will be active as relief worker teams:

- Police. For traffic control and public control. There are different kinds of policemen, for instance the national Dutch police force, the water police, and the harbour police. In this case they are considered as one, because their tasks are quite similar.
- Medical workers. In an emergency situation there is a chance that there will be victims who need immediate medical treatment. This will be done by ambulances with medical workers who are experienced in helping victims at emergency locations.
- Fire brigade. A fire in the port is a potential crisis. Firemen should try to control and diminish the fire as soon as possible.
- Commando Place Incident team (CoPI team). This is a team that is formulated at the RPA in emergency situations. The team consists of experts in different fields of experience. Together the team has knowledge about the local situation at the port. Because the team is going to help locally they can be considered as relief workers too.

Next to these relief workers there are some actors who support the relief workers directly.

- Experts. These experts are for instance in charge of measuring the amount of chemical gas in the air and giving an indication of the danger for all people and especially relief workers in the area.
- Rotterdam Port Authority (RPA). In case of a crisis these headquarters are occupied by employees from different departments: Traffic, Operations & Maintenance, Dangerous & Poisonous Substances, Security, Processing and Information, and Regulations & Law enforcement. The RPA is the first line of command when a crisis occurs. After the crisis has been reported, the emergency call centre contacts the RPA. The RPA is responsible for further emergency handling.
- Emergency call centre. The Emergency call centre is the first contact point in case of emergencies. The main role is to handle incoming calls and transfer the situation to the RPA. There is also a role for the emergency call centre to provide advice and information to incoming calls from people in the area.

Table 1-3 gives a general overview of the most likely activities for the different relief workers in case of a fire.

Table 1-3. Activities per actor in case of a fire

Actor	Activities
All relief workers	▪ Reporting status to call centre or headquarters ▪ Getting to accident location as soon as possible ▪ Sending information about accident once on location ▪ Finding the way to the accident location ▪ Knowing where the other (back-up) teams are
Policemen	▪ Managing traffic, for instance closing roads ▪ Managing public, keeping them away, evacuating ▪ Managing water traffic, stopping ships
Medical workers	▪ Bringing victims to hospital
Firemen	▪ Stopping fire ▪ Relief of victims from fire
CoPI	▪ Information gathering ▪ Local crisis management ▪ Informing relief workers

What is the kind of information those actors need, and what information can they provide? Table 1-4 answers these two questions.

Table 1-4. Different kind of information needed, and information provided

Actor	Information needed	Information provided
Relief workers	▪ Location of the accident ▪ Traffic information ▪ Danger of accident ▪ Number of people in area ▪ Detailed maps of the area ▪ Location and load of ships ▪ Chemical gas risks ▪ Location of relief worker teams ▪ Which actions should be taken	▪ Location of the accident ▪ Location of relief worker team ▪ Danger of accident

Actor	Information needed	Information provided
RPA and call centre	▪ Location of the accident ▪ Location of relief worker teams ▪ Danger of accident ▪ Number of people in area ▪ Detailed maps of the area ▪ Location and load of ships ▪ Chemical gas risks	▪ Which actions should be taken
Experts	▪ Location of accident ▪ Traffic information	▪ Chemical gas risks

The table shows that there is much more information required than the main actors can provide themselves. At least four resources of information are needed. To discover who can provide this missing information the following questions should be asked:

- Who provides detailed traffic information?
- Who has got information about the number of people in the disaster area?
- Who delivers detailed maps of the area?
- Who knows about the location and the load of the ships in the port?

1.7.2 Service concept

There is a need for a new service that enables relief workers to do their job better in crisis situations. We will call this service the Harbour Emergency Response System (HERS). The goal of the HERS is to facilitate relief workers with real-time and relevant information in crisis situations in the Port of Rotterdam. So in short: the main concept of the service that we want to design is a system that provides real-time information from different services and data sources about what the situation is and how to respond to it in emergency situations. This section will illustrate the service concept.

It is expected that better communication and better available information will improve the effectiveness of the relief workers. The service system will help by providing all kinds of information on the Personal Digital Assistants (PDA) of the relief workers. The relief workers can use the PDA to communicate with each other as well. On the screen, the first teams can indicate exactly where the disaster occurred, so other teams can take the shortest routes. If roads are congested or blocked, the teams can see that on their PDA and take another route to the disaster area.

So, the service system will provide five types of functionality to the relief workers. First, the PDA will provide the relief workers with information about the location of the disaster. Second, the PDA is used by the relief workers to communicate with each other. This communication can take place via voice and text exchange, but video and picture messaging will also be possible. The improved communication is expected to lead to higher productivity of the relief workers. Third, the PDA will show which actions need to be taken by the relief workers and which resources are available to them. This improves the quality and speed of decisions that are made by the relief workers. Fourth, the PDAs will store the information that has been exchanged and the actions that have been taken. As such, it enables ex-post evaluation about how the crisis was handled. Fifth, the service system will use the PDAs to find the location of the relief workers, which gives a better overview of the situation.

Figure 1-3 shows an example of a possible screen lay-out. A route indicates how to get to the fire as soon as possible. On the screen it is visible where relief teams are located. It is also possible to see which roads are blocked.

Figure 1-3. Screen-shot from a device with a map overview

1.7.3 Technical architecture

An overview of the technical architecture is given in Figure 1-4. The technical architecture is based on the three tier model (see chapter 3 of this book). This architecture consists of a presentation layer, an application layer and a data layer.

Data layer

Data is needed from various distributed sources. These include data from the currently used emergency response systems, data from a traffic information service, data about the location and loads of ships, and data about the working schedules of port workers. It is not desirable, and probably also not feasible to integrate all these data sources in the HERS service system. Therefore, all these existing data sources will stay in place, and all actors will remain responsible for their own data.

Application layer

The main task of the application layer will be to make sure that the right data comes from the right data sources to the right users at the right moment. This is also the main functionality of the HERS service system.

Presentation layer

The presentation layer describes how the service should be presented on the small screens and how the users can interact with the service system. For example, the location of other relief workers and of the disaster will be visualised on a map that is shown on the screen of the PDA. The relief workers can access a menu on their PDA in which they can scroll through different functions. Personal settings will be stored on the device and can easily be altered by the relief workers.

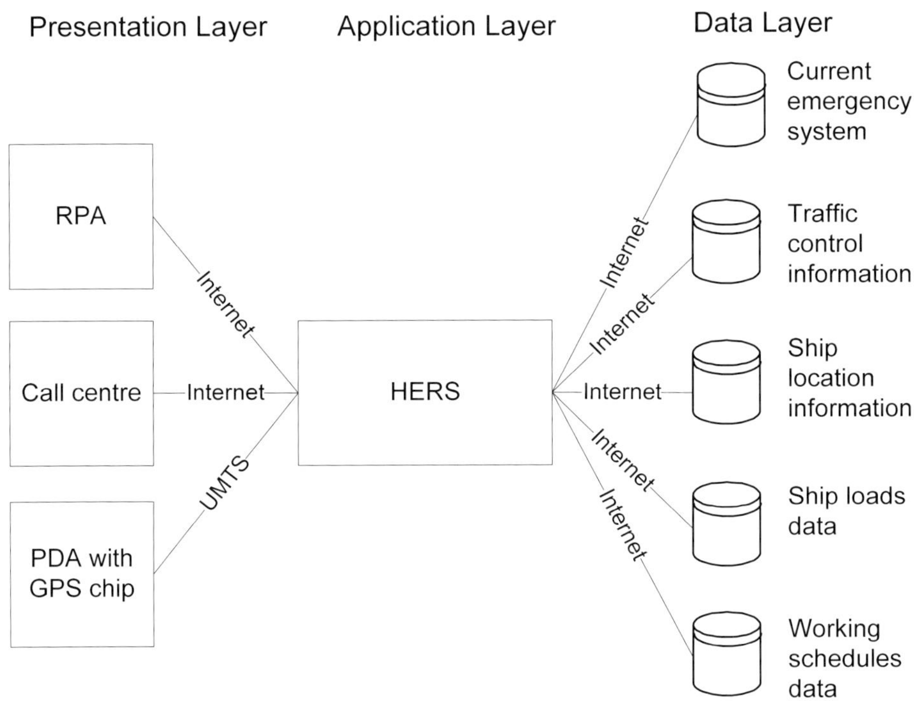

Figure 1-4. Technical architecture for the HERS service

The presentation layer also describes how the RPA and the call centre will receive information from the HERS service system. This information is, for example, about the location and status of the relief worker teams.

Communication

The communication between the data providing services and the application layer will be done by using, for example, the Simple Object Access Protocol (SOAP). This protocol provides a way to wrap information into 'envelopes' that can be exchanged platform independent between different services. A SOAP message from the traffic information service will for instance tell which roads are congested. Once this information comes in at the application layer it is immediately transferred to those devices who are receiving traffic information from that specific area.

Messages between the presentation layer and the application layer will be exchanged by using eXtended Mark-up Language (XML). This is a standard way of marking up data so the structure of the information is understood by sender and receiver. The PDAs will convert the received XML-data to visible information on the screen. Computers in call centre and RPA will do the same, though for larger screens.

Communication between the data providing services and the HERS service system will take place over the fixed broadband Internet lines that are already available. These lines are also used for communication between the HERS service system and the RPA and call centre. Communication between the application layer and the PDAs will take place over the wireless UMTS network. In order to enable UMTS communications some technical arrangements should be made with an UMTS service provider. The service management will pay the UMTS operator for the data traffic. A problem is the current low coverage of the UMTS network. In order to receive the signals everywhere in the port, a denser network will be needed.

To get information about the location of relief workers, some form of positioning technology is needed. Therefore, a GPS (Global Positioning System) receiver will be integrated into the PDA. It may also be possible to use the positioning capabilities of the UMTS network. In that case, a GPS receiver will not be needed. The coordinates of each device will be sent periodically to the HERS service system.

1.7.4 Result

This design example has illustrated how an existing service system can be improved. The design example description started by analysing the current organisational network and work practices of relief workers in the port of Rotterdam, and the shortcomings of these practices. Then, requirements for a new service were generated to solve these problems. Based on that, a new service concept was developed, as well as a technical architecture that supports this concept. Remarkable about the design example is that the organisational network remains the same with this new service system. This is an advantage, as changes in an organisational network can be time consuming and costly.

1.8 Comparison of the two design examples

Both designs pay attention to the service concept, the technical architecture, and the organisational network: one in the domain of Information and Entertainment and the other in the domain of Crisis Management. There are also other differences between the two design examples. One difference is that 'Guido the Guide' is aimed at consumers (b2c), whereas 'Support for relief workers in the Port of Rotterdam' is aimed at employees of the emergency services (b2b/b2e). Another difference is that 'Guido the Guide' is a design for a new service that does not replace any current services. The 'Support for relief workers in the Port of Rotterdam' design is aimed at enhancing the current working methods of the relief workers in case of emergency situations in the port. As a result, the design approach of the two groups was different, which can be seen from the section titles of the two design descriptions. The 'Relief workers' student team started by analysing the current way of working, the current problems, and the current organisational structure. Based on the analysis of the current situation, they formulated requirements for the service to be designed. After that, they designed the service, staying close to the current work practices and the current organisational network to enable the embedding of the service into the current organisational structure and work processes. The 'Guido the Guide' group designed a totally new service that did not build on existing services that could be studied. Therefore, they started with a service concept without analysing the current situation.

1.9 Structure of the book

The structure of the book is as follows. In this introductory chapter, we have explained the *domain* of (mobile) service systems. In the next three chapters of the book, the *aspects* of the domain will be described in detail; this part answers the question "*what is* service systems design"? The chapters pay attention to the following aspects: service concept (chapter 2), technical architecture (chapter 3), and organisational network (chapter 4). Within and between the chapters, practical examples are given to illustrate the theory presented in the chapters. After this, we explain *how to design* those systems. Methodologies for service systems design should be able to support the design process for both existing services and new services. Chapter 5 provides the methodological background, and chapter 6 gives an example of a design method that can be used for designing mobile service systems. Chapter 7 provides an extensive case study with an application of the design method of chapter 6.

2 Service Concept

The service concept is the differentiating value proposition as demanded by the end-user. It expresses how the service creates value for users. The 'service concept' is one of the three aspects of the service systems framework, and it will be further explained in this chapter. First we discuss services in general and information services in particular, then we will zoom in on mobile information services. We describe ways to categorise mobile information services and present the characteristics of mobile information services; divided into general characteristics and characteristics that are specific for the consumer and business market. Next, we discuss the value and the quality of mobile services. This is followed by theories around the acceptance of services and pricing of services. The service concept aspect of the service systems framework is illustrated with two examples of mobile services, one for consumers and one for businesses.

2.1 Concepts of mobile information services

Designing services differs from designing physical goods since services and goods have different characteristics. The services literature provides a set of generally accepted definitions of traditional services (Bateson, 1992; Grönroos, 2001; Kasper, van Helsdingen, & de Vries, 1999; Kotler, 1997; Zeithaml et al., 2006). Grönroos identifies three basis characteristics of services: 1) services are processes consisting of activities or a series of activities rather than things; 2) services are at least to some extent produced and consumed simultaneously; 3) and the customer, to a greater or lesser extent, participates in the service production process. A comparison of a set of similar and generally accepted characteristics of traditional services, and the equivalent characteristics of physical goods is presented in Table 2-1.

Table 2-1. Differences between services and physical goods (Grönroos, 2001)

Services	Physical goods
▪ Intangible	▪ Tangible
▪ Heterogeneous	▪ Homogeneous
▪ Production, distribution and consumption simultaneous process	▪ Production and distribution separated from consumption
▪ An activity or process	▪ A thing
▪ Core value produced in buyer-seller interactions	▪ Core value produced in factory
▪ Customers participate in production	▪ Customers do not (normally) participate in production
▪ Cannot be kept in stock	▪ Can be kept in stock
▪ No transfer of ownership	▪ Transfer of ownership

The differences between products and services (as shown in Table 2-1) have consequences for adapting product design theories for service design. E.g. the quality of an activity or process is more difficult to measure than the quality of a thing. The question is whether a mobile information service can be considered as a pure service, because the mobile service is so dependent on a handset. The characteristics of the handset are relevant for designing mobile information services, e.g. the size and resolution of the screen for the user interface of the service, and the keyboard of the mobile phone with limited data entry possibilities. The services and the physical goods necessary to use the service are related and interdependent. In addition to the differences between services and physical goods, electronic services differ from traditional services. The common denominator that distinguishes e-services from regular services is the availability of the service via an Internet channel, which makes it accessible from anywhere and consumable 24 hours a day. The following definition for an e-service is based on the definitions of Grönroos (2001) and Kasper et al. (1999):

'An e-service is an activity or series of activities of intangible nature that take place in interaction through an Internet channel between customers and service employees or systems of the service provider, which are provided as solutions to customer problems, add value and create customer satisfaction'.

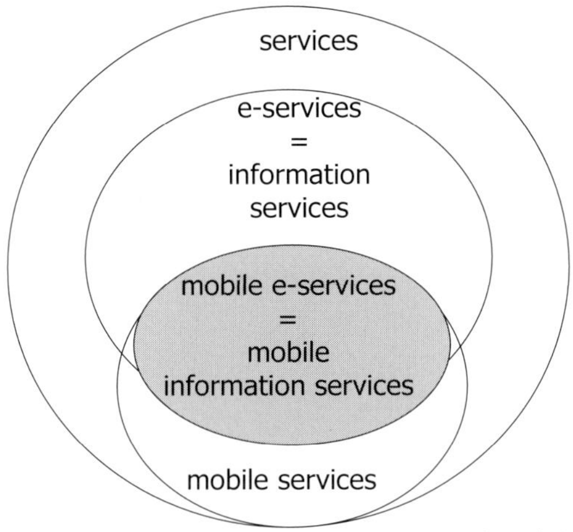

Figure 2-1. Positioning services, e-services and mobile e-services

Mobile e-services are a subset of both mobile services and of information services or e-services, as shown in Figure 2-1. Mobile voice services are a subset of mobile services but not of mobile e-services. Considering the dynamic developments around technology and the use of technology one might wonder whether the distinction between e-services and mobile e-services will still exist in the future. After all, mobility keeps growing and people request the information they need more and more place and time independent.

When the information services definition is extended to include the mobile aspect the following definition for mobile information service is derived:

> *'A mobile information service is an activity or series of activities of intangible nature that occur when the consumer is mobile and a mobile telecommunications network supports the interaction through an Internet channel between customers and service employees or systems of a service provider. The aim is to provide solutions to customer problems and needs, add value, and thus create customer satisfaction'.*

The heterogeneous quality of mobile e-services and the difficulty of measuring their quality are relevant when designing mobile information services. Given this heterogeneous quality, the design of mobile e-services is complicated by the fact that different stakeholders are involved in the design of a mobile information service; that all these actors may have different perceptions regarding the user requirements for the services; and that these expectations regarding the technological performance often exceed what is possible in reality.

A widely acknowledged way of looking at service quality is to make a distinction between the outcome (what) and the process (how) (Grönroos, 2001). The service outcome quality, also defined as technical quality, can be measured objectively by both producer and consumer.

The 'what' represents the value of the service for the users. We further discuss the 'what' in section 2.3; the 'how' is related to the process and how the user perceives the quality of the service. The processes determining the added value are discussed in section 2.4, service acceptance in section 2.5, and the service quality dimensions in section 2.6. First, different ways to categorise mobile services are described.

2.2 Categorisation of mobile services

Mobile information services are a merger of Internet and mobile telecommunication. Unlike the use of the mobile phone for communication services like making calls and sending SMS, mobile information services are not widespread in the market.

There are many ways to categorise mobile information services. Mobile services can be classified by their basic function: information, communication, transaction and entertainment. All these basic functions might make use of the possibility to position the user. The subset of mobile information services that makes use of the positioning of the user is called location-based mobile services; often abbreviated as LBS (location based services). In subsection 2.2.1 we will further discuss characteristics of location-based services. First, we present three other approaches to categorise mobile services, in each category location-based and non location-based mobile services can be found. Examples are provided in subsection 2.2.2.

Applications

A very general mobile application is mobile commerce. Mobile commerce is the term for the extension of electronic commerce from wired to wireless computers and telecommunications, and from fixed locations to anytime, anywhere and anyone (Keen & Mackintosh, 2001). Turban et al. (2004) define mobile commerce as any e-commerce done in a wireless environment, especially via the Internet. Mobile commerce can also be viewed as mobile services with a transaction function; customers buy something from a third party using a mobile device and mobile telecommunication as a channel. By transaction services we mean services that facilitate the payment between third parties and customers for e.g. wireless shopping or mobile ticketing.

Varshney and Vetter (2001) label 'value-added services in mobile commerce' as applications and classify them as follows: mobile financial application, mobile advertising, mobile inventory management, proactive service management, product locating and shopping, wireless re-engineering, mobile auction or reverse auction, mobile entertainment services, mobile office, mobile distance education, wireless data centre and mobile music. You can use mobile services anyplace, anytime, real time and with positioning you can determine your own location, and use the location of other persons or objects. The technology to enable localisation is available and develops continuously. Barnes (2003b) categorises location-based mobile applications in four key areas: Safety, Navigation or Tracking, Transactions, and Information. Giaglis et al. (2003) provide a taxonomy of location-based mobile applications in which they distinguish the use for: Emergency, Navigation, Information, Advertising, Tracking, and Billing. Different applications pose different requirements for the accuracy of the location information and speed of the network. Other requirements are also affected by the type of application; trust in terms of privacy and security will be very different when designing a service for 'mobile shopping' or for 'mobile music'.

There is also a distinction between *wireless* and *mobile* applications. Kalakota and Robinson (2002) state that wireless commonly means mobile and online. This means that there is a real-time live Internet connection via satellite, cellular or radio transmitters. Mobile and offline means that the device runs self-contained programs but is not connected to the Internet, for example a synchronised PDA. Sabat (2002) uses a matrix to categorise emerging solutions by the access mode (fixed or mobile) and the mode of connection (wired or wireless) of the device with the content source in the wired world. Broadband wireless access is an example of wireless-fixed; and the mobile Internet is an example of wireless-mobile. This distinction is important because it affects the way services are designed and used.

Service model

Services might be pushed or pulled. Services in the 'pull category' enable users to 'pull' information wherever and whenever required, such as travel directions, taxi hailing, mobile yellow pages and m-commerce. Services in the 'push category' utilise the position of the mobile device to determine whether the user meets the criteria of a potential customer or service recipient. An example is a mobile advertisement based on personal characteristics of the user, e.g. his or her location. An 'in between' service model is a model where the user first subscribes him/herself for a service and allows the service provider to use his/her characteristics to push specific information. Privacy is an important issue for push services.

Market segment

Services might be targeted at the consumer market, business market, or public sector. Employees can be separately regarded or as part of the business market. Market segments can be directed to consumers (b2c, e2c, g2c or c2c) and to businesses including employees and the public sector (b2b, b2g, g2b or b2e)[1]. Consumers might also be categorised by how they adopt services. Users might be divided by their adoption behaviour of technologies as described in Rogers' Bell-curve in which the market is split up into innovators, early adopters, early majority, late adopters and laggards (Rogers, 2003). Another way of looking at segments is the demographic method, looking at age, sex, income, area, etcetera.

2.2.1 Location based services (LBS)

Making a distinction between location-based services and non-location based services is another way of categorizing mobile services. However, positioning is an inherent feature of mobile services that offers many opportunities for value proposition to customers. It is expected that in the future the distinction between mobile services and location-based mobile services will vanish when this feature will be optimally exploited. Location based mobile services are attractive to both consumers and businesses alike and offer safety, convenience and productivity (Turban et al., 2004). If a third party wants to offer services to the end-user, an ability to locate this customer may help the supplier to deliver services that are customised to his needs in a specific spatial context. The European Telecommunications Standards Institute (ETSI, 2001) describes location based services as a network-provided enabling technology consisting of standardised service capabilities enabling the provision of local applications. This term 'location based services' is used alternately with the term 'location-based mobile services' (LBMS). The latter term indicates the type of service more precisely. LBMS can also be interpreted as any service in a given location, e.g. delivering groceries to the home, enabled by a fixed infrastructure or by digital traffic information. We define location-based mobile services as services based on the mobile telecommunication network, accessible by the mobile

---

[1] in the x2y acronyms, b stands for business, c for consumer or citizen, e for employee, and g for government.

device of an end-user, and making use of the automatically determined location of the mobile station. A more practical shorter definition is: location-based mobile services are services that exploit information about where a device (mobile user) is located.

According to a white paper by Gravitate, Inc. in Koeppel (2000), different generations of location-based mobile services can be distinguished. The first generation of services relies on the user to provide the location information manually – e.g. as a street address or postal code – to applications, and the application returns maps and routing information. The next generation of services has some built-in positioning ability that is sufficient to provide routing or proximity information, for example, to find the nearest restaurant or petrol station. The third generation of location-based services has greater position awareness and thus can also deliver services in the push service category, such as notification of events or traffic alerts. A fourth generation of location-based mobile services consists of location-aware applications based on infrastructures, technologies and techniques that enable context aware information to be seamlessly offered to the end-customer.

Localisation should be an integral part of context aware services. The development of context aware services is in its early days and there are many technology related and customer behavioural issues that need to be further explored. In the long term, location-based mobile services are likely to be integrated to offer customised, localised content wherever and whenever necessary and applicable.

2.2.2 Examples

Examples of mobile services in which the environment of the user is important, are given in Table 2-2 (based on various literature sources). The services in Table 2-2 are categorised in two ways. The heading row indicates the market segment and the first column indicates the type of application.

Table 2-2. Examples of (context aware) mobile services

	Consumer Market (b2c, e2c, g2c or c2c)	**Business Market (b2b, b2g, g2b or b2e)**
Information	▪ Geographic messaging of general, local information like news and weather at specific location ▪ Information update on public transportation and traffic ▪ Yellow pages ▪ Product information; track the product life cycle in the shop ▪ Shopping locator services ▪ Financial services like foreign currency info, closest ATM	▪ Business travellers services like scheduling and local traffic information ▪ Geographic messaging e.g. when you pass reception of company or other relevant location. ▪ Financial services: foreign currency info, closest ATM. ▪ Yellow pages ▪ Mobile workers: planning, hour-registration, invoice information, kilometre expense account. ▪ Vehicle tracking ▪ Location-based scheduling and tracking for sales forces and field-service engineers; ▪ Customer Relationship Management tools to mobile workforce
Entertainment	▪ Guided tours with hotel, restaurant, pubs, cinema, theatre and events information ▪ Locate and direct people for dating and friend-finder services ▪ Location-based gaming ▪ Mobile (interactive) TV ▪ Video clips	▪ Business relationship services; trips, hotels etc.; and services like translations and guided tours
Navigation and tracking	▪ Navigation applications: best route from A to B; driving directions ▪ Tracking service: find persons and/or goods, children, stranded drivers. ▪ Alerting services	▪ Supply chain management, Inventory management, Fleet management ▪ Tracking and tracing of assets like transported goods

	Consumer Market (b2c, e2c, g2c or c2c)	Business Market (b2b, b2g, g2b or b2e)
Transaction	▪ Location-based billing e.g. for automatic pricing of road usage ▪ Wireless shopping assisting services for cross selling ▪ Public transportation reservations ▪ M-Tickets: ticket purchases, i.e. for movies, theatres and concert ▪ Buying local content ▪ Mobile Coupons, i.e. discounts	▪ Location-based billing ▪ Tax collection dependent on localisation e.g. mileage registration ▪ Just In Time orders from places without wired infrastructure, e.g. construction sites
Safety	▪ Tracing services: tracing mobile terminals to provide safety i.e. child and elderly people tracing; prevent theft by car tracing ▪ Emergency services ▪ Roadside assistance	▪ Security and access services ▪ Hazardous goods transport ▪ Localisation of guards

Bundling of services focussed on the wants and needs of an individual offers added value for this individual. The integration of services around both business and personal use drives real economic value in mobile commerce (Malhotra & Segars, 2005). Nevertheless, some services will be targeted on the consumer market and some on the business market.

2.3 Mobile information services and their value

The value of a service is mostly measured by the specific benefits for the context in which the service is delivered: "Is the service useful?" It is difficult to establish uniquely whether a service is useful to a customer, as we can find various drivers like saving time, easier execution of processes, improved quality, working more quickly, better job performance, increased productivity, and higher effectiveness (Davis, 1989). These drivers constitute the value proposition of the service.

The value of mobile services has been considered in recent research projects and publications on mobile services. The value of mobile services arises from the freedom that they offer. Nah et al. (2005) referring to

(Keeney, 1999) define the value propositions of mobile applications as the net value of the benefits and costs associated with the adoption and adaptation of mobile applications. The value mobile services offer is directly related to the characteristics of the mobile application(s) that provide access to the service. These applications are the interface to the service. Clarke (2001) mentions that the functionality of mobile services is based on the new value areas of ubiquity, localisation and convenience.

Baldi and Thaung (2002) mention as distinctive characteristics of mobile services ubiquity, accessibility, reachability, localisation and personalisation. Anckar and D'Incau (2002) identify the following five different settings in which mobile services can create value: time-critical arrangements; spontaneous decisions and needs; entertainment needs; efficiency ambitions; and mobile situations. In these settings people might adopt mobile information services. The adoption of new technology further depends on factors such as perceived usefulness, perceived ease-of-use and other behavioural beliefs and normative beliefs. This is further explained in section 2.5

For mobile services it holds that the value they might offer depends on the context of the user. The different settings described above are part of this context. For example, is the user a tourist who visits an unknown city? Is the user a victim in a crisis situation, such as a car crash where the driver has no idea of his or her exact location or is he/she a parent whose daughter is missing at Disneyworld, or an elderly citizen living alone who suddenly faints? Context should be the starting point in the service design process and the requirements resulting from the context analysis have to be matched with the capabilities of the available technological components (Blom, Chipchase, & Lehikoinen, 2005; Tamminen, Oulasvirta, Toiskallio, & Kankainen, 2004).

A mobile information service can only be context aware if there is enough relevant content. In this regard, however, it is also important to manage the expectations of the customer. The *Mobile* Internet is not the Mobile *Internet*. The richness of the content on Internet is just not available on the mobile phone. Of course, it is not desirable that the complete content of the *Internet* is available on the mobile device. People on the move do not want to read large documents on a mobile device. The mobile feature means that rich content should be available that is adapted to the context of the user.

2.4 Processes determining the added value

A service system is aimed at providing added value for its users. This added value is realised when users can use the service system, and really do use the service system. Figure 2-2 shows three different processes that can be distinguished in the design of a service system and which influence the added value for the user.

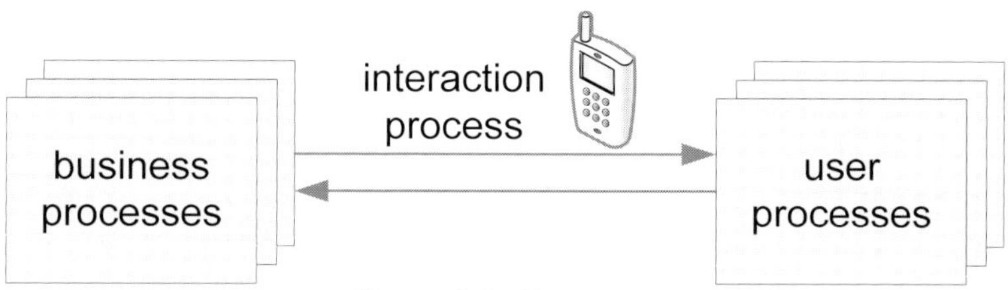

Figure 2-2. Processes

At the right hand side of Figure 1-2 the *user process* is depicted. This process is central in the design of a user-centred service system. The service system designer has to be aware of the context in which the user uses the system. What logical steps from the user's standpoint of view can be distinguished for using the service? Are there different situations in which the user might want to use the service? Is the user process with the service system usually carried out under time pressure? If so, how can the system help maximally to make it easier for the user to reach the required state quickly? Many services fail because the way the service is offered does not match the way the user wants to use the service. An example could be a sequence of steps in the service delivery which is not logical to the user. When designing a good service, the following aspects need to be analysed or designed: the roles of the users, the contexts in which the users use the service, the relation with other users on whom the service delivery might be depending, and the sequence of user activities that the service system needs to support. When deploying a service in an existing situation, users might already have a preferred way of working that is not easily mapped on the service system. In that case, activities of users might have to be redesigned in order to create a successful service.

The second process is the *interaction process*, which links the user processes and the business processes by the use of a (mobile) device. This interaction process has to be carefully designed as well, because the linkage of the user and business process is not easy, especially when taking into account different preferences from different users in different contexts. When using a hand-held device, the requirements for the interaction process are even more difficult, because the device has a small screen and a small keyboard, or even just a numerical keyboard in the case of a mobile phone. The aspect of *ease of use* of *usability* of a service is therefore one of the most important success or failure factors in service systems design.

In the *business process* the service delivery is prepared for the user. This process consists of reacting to user requests, gathering the right data, and preparing the presentation to the user. In the background, several services can interact, for instance when some of the data has to be gathered from an external source. In the application, invisible for the user, the user's preferences, and information on characteristics of the communication device are stored, so the service can be delivered in a personalised way that fits the device used. These processes have to be designed in such a way that the response times are acceptable for the user, and that the service does not stop working completely when links to external services or external data sources are temporarily unavailable. The business processes have to be designed in such a way that the service delivery satisfies a well-defined service level. A service for emergency workers will of course have a much higher minimum service level than a tourist information service.

When designing service systems for b2b situations, the business processes and the user processes are tightly linked. This makes it even more difficult to design the operational processes within a business. Take for example an insurance expert who visits customers to sell new insurance products or to change current insurance policies. The mobile services to support these processes have to match both the personal style of the salesperson, and the rigid structure of the information systems of the insurance company. Moreover, the salesperson is part of the insurance information systems, and in the design of these systems, the sales process and policy maintenance processes have to be taken into account. As a result new procedures might have to be implemented. When the system is extended with mobile services to work with insurance policies at the customer's location, this might not match with the approval procedures in the company. In the old situation, an approval procedure was followed that could take several days, whereas the salesperson now wants approval within 30 seconds. Clearly, the business processes and the design of the new service are not yet aligned to deliver the service according to the wishes of the salesperson. Zeithaml et al. (2006) explain the difference of managing service quality by a gap model. They distinguish the following five gaps between the: (1) customer's expected service and marketer's perception of customer expectations; (2) marketer's perception of customer expectations and the translation of these perceptions into service quality specs; (3) translation of these perceptions into service quality specs and actual service delivery; (4) service delivery and external communication to customers and (5) the customer's perceived service and expected service. Before we look further into service quality we first explore the concept of service acceptance.

2.5 Service Acceptance

The consumer to a greater or lesser extent, participates in the service production process. Via the user interface, the user interacts with the provider and in that manner he/she 'consumes' the service. The user interface consists of all components that allow the user to interact with the service system; this may include a mobile phone, a PDA, an office PC, or something else. In this book however, we focus on mobile services: What does the process of 'consuming' the service look like in this context? And what aspects are relevant in designing this process?

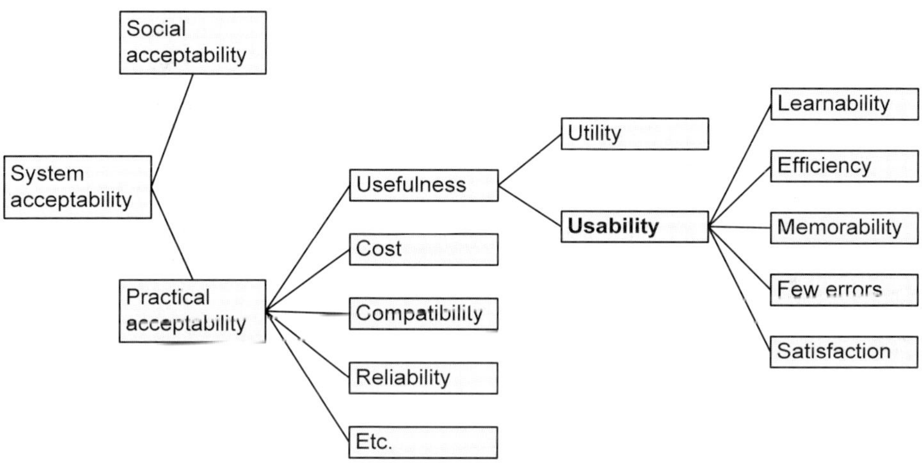

Figure 2-3. System acceptability (based on Nielsen, 1993)

To formulate an answer to these questions we will use the insights of system acceptance models. These models provide an overview of the issues that influence the acceptability of a system. The first model we want to discuss here is the model presented by Nielsen (1993), see Figure 2-3. According to Nielsen system acceptability is a combination of social acceptability and practical acceptability. The social context of the user is also relevant; social aspects also influence the adoption and manner of using the service. In the model of Nielsen this aspect is depicted with the concept of social acceptability. A close analysis of the context in which the activity takes place is needed to understand why this activity is performed in a certain way and how tools can be adapted or designed to support it. This can be shown by an example: imagine you want to develop a mobile device that can be used in the ambulance to contact a specialist at the hospital. This device may be very useful in a case where a patient is injured in such a way that advice from this specialist supports a more adequate treatment. In this example it is not useful to focus only on the nurse in the ambulance. His main activity, treating the patient outside the hospital, contributes to the main outcome of providing the patient with medical care. Other people of the community contribute to this main goal as well. Implementation of the mobile device has a high impact on the division of labour and rules applied in this community (Example based on Van den Anker, 2002).

Practical acceptability is a result of the combination of characteristics of the system, i.e. usefulness, cost, compatibility, reliability, and other factors. Where *usefulness* is related to how suitable a system is in achieving a certain goal. This concept is a combination of the utility and the usability of a system. *Utility* describes whether appropriate functionality is at hand and *usability* is defined as the extent to which a product can be used by specified users to achieve specified goals with effectiveness, efficiency and satisfaction in a specified context of use. In this last definition *effectiveness* refers to whether users can reach stated goals and *efficiency* relates to the resources needed to reach these goals. Nielsen further defines a usable system as a system that meets the criteria of learnability, efficiency, memorability, few errors, and satisfaction. These criteria are discussed in more detail in section 2.6.2.

A slightly different approach is presented by Davis (1989; 1993). He describes that users are likely to use a system when the system is perceived as useful and perceived as easy to use (see Figure 2-4). In his model Perceived Usefulness (PU) is defined as "*the degree to which a person believes that using a particular system would enhance his or her job performance*" and Perceived Ease of Use (PEOU) as "*the degree to which the person believes that using a particular system would be free of effort*" (Davis, 1989, p. 320). This effort consists of physical effort, mental effort, and ease of learning the system. The two concepts Perceived Usefulness and Perceived Ease of Use both influence the decision of (intended) users to use technological tools. In addition to his model Davis presents a questionnaire that can be used to measure the concepts in the model. This questionnaire enabled the application of the TAM model in several studies. Although the TAM model originally focuses on the adoption of technology in organisational settings, from these studies it appeared to be applicable in other settings as well.

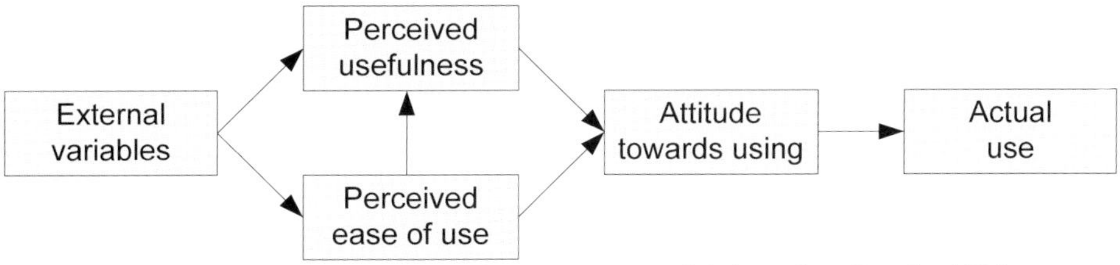

Figure 2-4. The Technology Acceptance Model (based on Davis, 1993)

Compared to the model presented by Nielsen (1993) the TAM model is more generic. However, Perceived Usefulness in the TAM model is comparable with the utility concept in the model of Nielsen. And, the concept of Perceived Ease of Use can be related to the usability concept in the model of Nielsen. The TAM model further refers to external variables that influence this attitude towards using the system, whereas Nielsen specifies some of these external factors. We will use the model of Nielsen as it appears to provide more handles in understanding the use of systems.

How can the concepts presented by Nielsen (1993) be applied to service systems? From his overview we learn that if a service (system) is considered useful, it is more likely to be accepted and used. By translating the concept of usefulness we assume that a user should consider the service as having utility and being usable. In other words, the user is likely to consider a service useful when it fits its needs and tasks and when the service is easy to use. The two concepts of usability and utility are highly interrelated. However: a user interface that is usable may contribute to a service being perceived as having the utility to provide appropriate functionality. Vice versa, if a service (theoretically) has the utility to provide appropriate functionality, but it can only be 'consumed' via a badly designed user interface, a user may avoid using the service. The user might not be able to see through this interface and as a result evaluate the service as not useful. Or, he/she may not want to put effort into overcoming the difficulties of the interface.

Trade-off perceived usefulness and switching costs

Briggs et al. (2001) translate perceived usefulness into perceived-magnitude-of-net-value and the perceived-frequency-of-net-value. The perceived-magnitude-of-net-value, refers to "an attitude, a balanced subjective assessment of the probable consequences of changing from existing technology to the proposed technology. Upon being exposed to the technology, prospective users will synthesise a holistic sense of how their lives will be different if they change to the new technology." The second concept, the perceived-frequency-of-net-value, refers to the frequency users expect to derive the net-value they perceive. These concepts together influence the likeliness users will adopt a new system. As a third concept they introduce the moderating variable of perceived-net-value-of-transaction. This concept presents the value derived from the transition activity self; it represents the balance estimated by the user, between switching-costs and switching-benefits. This value is not only relevant as a mediator in case of replacing technologies, but also in case of newly introduced technologies. In the context of service systems: is the service attractive although it requires some time learning to use it?

Trade-off perceived added value and "real" added value

There can be a major difference between the perception of the user about the added value and the "real" added value that can be established with, for instance, an economic model. Especially in b2c, g2c, and c2c services, the users will *select* the service on the basis of the perceived added value, and if that is judged to be low, users will not start using the service, even if the real added value is high. Therefore, it is important to pay attention to the perception of the users, through a good service concept and a well-defined user interface, and to the 'fit' between the activities of the user and the sequence of activities as offered by the service system.

Pressure to use a service

In the success of mobile services the pressure to use a service is another important factor. For some processes, use may be forced. It is not possible to neglect a service that is part of an organisational process that one has to contribute to.

Another type of pressure that may be experienced by potential users of a service is social pressure: the pressure of other users using a specific service. If someone feels they are the only one not using the application, and he or she wants to fit into the group, they might feel as though they should also start to use the application. The adoption of a service may be partly determined by the social pressure of colleagues for example.

Critical mass

Of course, the success and usage of mobile services also depends on the number of users. Adoption likelihood is sensitive to critical mass and anticipation of future network size (Briggs et al., 2001; Shapiro & Varian, 1999). The aspect of critical mass may influence the success of a service in two ways: the first we will refer to as economical sustainability and the second service content sustainability. The first principal, economical sustainability refers to the need of developing a profitable service. Profitability is a prerequisite for suppliers of information or other necessary contributions, to continue to invest in the service. This implies, for example, keeping the information up-to-date. A mobile navigation support application looses its usefulness if no new updates of the map information are released.

The second principal, service content sustainability, stresses the relevance of critical mass in the case of a mobile service supports interaction between two (or more) parties. Fax and telephone are good examples of this aspect of critical mass. A fax is only useful if someone else has a fax too. And, the more people own a fax, the more useful your fax becomes. A 'friend finder' service only makes sense if enough 'friends' are also using the service.

2.6 Service Quality

The service process quality cannot be measured as objectively as the elements of technical quality because the customers' presence affects the delivery process. Quality regarding the production process is widely defined as meeting customers' expectations. The best known service quality dimensions are proposed, and improved in later versions, in the SERVQUAL model of Parasuraman et al. (Parasuraman, Berry, & Zeithaml, 1991; Parasuraman, Zeithaml, & Berry, 1985 1988): i.e. tangibility, responsiveness, reliability, assurance and empathy. Liljander et al. (2002) published quality dimensions of 'pure' e-services based on the dimensions of the SERVQUAL model. We describe them here more specifically for mobile information services.

- *Reliability:* currency and accuracy of product information, correct technical functioning of e-services, explanation of purpose and accuracy of execution of service promises. This is very relevant for mobile services considering the fact that users are on the move and might be in time-critical situations.
- *Responsiveness:* providing prompt service, help available, directly providing the information the customer wants to address his/her inquiry; both the dependency on the network and potential time-critical situations enforce this requirement.
- *User interface design:* overall design, ease of navigation, finding needed functionality without difficulty, overall ease of use and aesthetics. The small screen of mobile devices poses special requirements on the user interface.
- *Trust:* feeling of confidence and trust regarding privacy and security when using e-services.

- *Customisation and personalisation:* presenting the information in a flexible, adequate and tailored format. This can be done both by and for the customer. We define personalisation as a certain type of customisation and therefore as a category within customisation. For instance, mobile users personalise the screen, tunes and navigation menu of their mobile phone. Mobile services can usually be customised to avoid the need to enter personal preferences with every use of the system.

2.6.1 Reliability and Responsiveness

The infrastructure and the environment of the user influence the reliability and responsiveness during the usage of the service (Van de Kar, Wang, & Muniafu, 2006). To derive reliability, the service system should be connected with different communications and data networks that ensure high availability of services. The services need to be decoupled from the infrastructure, to handle failure issues and provide ubiquitous access to services. Furthermore, before engineers start to assemble service components, the separate components have to be proved to be reliable. The required robustness of the service also depends on the context of usage. The required robustness is less for a tourist who requires information than for a policeman in a crisis situation.

The service should handle the changing use of the services in the changing contexts of the user. The context study is important for this. Also changing infrastructure circumstances require that the service system should be flexible enough to handle changing user needs such as during periods of limited bandwidth, loss of signal, etc. and ensure that the service can still be delivered even at diminishing levels.

2.6.2 User Interface Design

The interaction processes refer to the way the user interacts with the service. Relevant aspects of this process concern the available devices to interact with the service, the operation of these devices, and the presentation of information on these devices. As most services are used via mobile devices, the limitations of these devices affect the overall performance of the interface. The device has to be small and easy to transport for example and as a result screen size is limited: the portability versus usability trade-off of mobile devices. To meet new prerequisites and requirements in design for these types of devices a number of new interaction techniques have evolved. Consequently they offer a number of interaction channels for input and output (Johansson, Lind, & Sandblad, 2005). As shown in section 2.5, Nielsen

(1993) defines five aspects that contribute to a user friendly or usable user interface. These will be discussed below.

Ease of learning

The system should be easy to learn so that the user can rapidly start getting some work done with the system. There is, for example, a big difference between user interfaces that are intuitive to use, because the interface is tuned to the terminology of the user, and user interfaces that are 'generic' and for which the user always has to make a translation between the terms of the user interface and his or her own terminology.

Efficiency of use

The system should be efficient to use, so that once the user has learned to use the system, a high level of productivity is possible. Examples of measures that can increase the efficiency of use are: storage of previous decisions, automatic filling of fields for the user, and following a workflow that is close to the user's workflow. Ease of learning and efficiency of use can be conflicting. In order to make a service system easy to learn, extra help can be given to the user, which an advanced user might resent because it delays efficient use of the system. Some systems offer the possibility to choose between a 'novice' mode and an 'expert' mode, where the latter mode is much faster as it omits all kinds of explanatory information and in-between steps.

Memorability

The system should be easy to remember, so that the casual user is able to return to the system after some period of not having used it, without having to learn everything all over again. Memorability is different from ease of learning in the sense that there is a difference between first-time use, and using the system again after a while.

Errors

The system should have a low error rate, so that users make few errors during the use of the system, and so that if they do make errors they can easily recover from them. Furthermore, catastrophic errors must not occur. (See also Lewis & Norman, 1986 on errors and error prevention). Rolling back to a previous state of the system (the undo function in many programmes) should always be possible. When an action cannot be undone, a user should be warned about it.

Satisfaction

The system should be pleasant to use, so that users are subjectively satisfied when using it; they like it.

2.6.3 Trust

People only want to use services if they trust what the effect of the service will be. Trust is a subjective matter, it is about feeling confident. Trust is a much researched item in the context of the Internet. It is closely related to security. "Security has to be one of the design criteria pervasive in system and service engineering. The way security is included has strong implications for functionality, efficiency and costs. The trustworthiness of the service depends on adequate provisions of security throughout in the chain. The converging sectors of the industry (including areas like internet and IT, infotainment, mobile communications, consumer electronics, etc.) carry a different history on security perspective and technical solutions. Therefore, security should integrate and evolve the existing security technologies through open standards in order to meet the requirements of the future advanced service products. The great challenge for the future security framework is to maintain simplicity and efficiency without leading to heavy implementation solutions. Ultimately, it is the user experience that drives the adoption of new services. The level of offered security should adapt the service needs in terms of user authentication, information encryption, privacy, anonymity, identity management, and content delivery. Security demanding applications like m-payments, m-health, content distribution, user profile management, are expected to drive the evolution towards a future security framework" (Tafazolli et al., 2005).

Security issues also have implications for the relevant regulatory and legislation framework. E.g. copyrights; anti SPAM regulations; privacy regulations; etceteras.

Privacy and security can be considered as opposite values. Sometimes people take less privacy for granted to gain more security. They know for instance that their position might be known by authorities, but by allowing this they can always be found in case of accidents or attacks. Privacy is also an issue related to push advertisement services.

In section 2.2 it is mentioned that one way to categorise services is by their service model. This service model might be pull or push or an in-between method. These methods are also called delivery methods. The delivery of pull based services is triggered by the user him/herself who searches for information based on his/her location. Thus, the location information of the user is used for the service. For push based services the delivery of the service is triggered by a third party and not by the user him/herself. For these services privacy is an issue. Privacy sensitivity is context dependent and changes according to the intensity of media reaction and focus on specific "cases". Customers should have the option to not share their context information, and they should not be damaged or harmed by their context or location information becoming known.

Dealing with privacy is possible on a so-called opt-in or opt-out basis. The first means that the user first gives permission to use his/her data. Opt-out means that the user's data are used until the user writes a letter with a request to delete his/her data from the mailing list.

There is both governmental regulation and self-regulation within the industry dealing with anti SPAM regulation and privacy regulation. Anti SPAM regulation might be applicable for pushing advertisements without an active request from the end-user. Camponovo and Cerutti (2004) have researched the anti SPAM regulations in several countries and concluded that within the European Union an opt-in principal is obliged by the regulations. An example of self-regulation is the code of conduct that is introduced for location based services in the UK. This code of conduct describes rules for child location services, friend locations services and mobile gaming services that use localisation (http://www.spiked-online.com/pdf/ LocationServicesCodeofPractice.pdf).

Safety is also important in a completely different sense: the safety that the usage of mobile services offers. For people driving by car (or bike!) in the dark it is a real necessity to have a mobile phone with them in case the car breaks down or they are harassed. Having a mobile phone might make the difference between staying at home and going out.

2.6.4 Customisation and personalisation

In order to make service systems successful, services have to be customised and personalised. Lots of unnecessary interactions with the user, such as browsing long lists within applications, can be avoided when the device knows where the user is, and what the most likely services are that the user would want at that location and that point in time. When context aware technologies have learning capabilities, context aware services can quickly adapt to specific user needs, again decreasing the amount of interactions needed. Imagine a business traveller who is in a meeting that takes much longer than originally planned. When the interdependencies between the used services are known, business services can immediately deduce that several bookings need to be cancelled, shifted, or changed when one of the services (for instance a plane booking) is changed. When the business processes also recognise the preferences of the user, a new chain of potential bookings can be created, and the traveller can be provided with one or two suggestions for new itineraries, probably within a few seconds or minutes. Although the implementation of location awareness and personalisation seems easy, it is extremely difficult. The context of a user consists of an unlimited number of variables, and the 'right' variables need to be identified that determine the success of the deployment of mobile services in practice. Even 'location' has many dimensions: when someone is in a business meeting, 'location' could be the address of the meeting location, its GPS location, the building in which the meeting takes place, the room in which the meeting takes place, or the business meeting itself. Depending on the service, each of these contexts might contain valuable information for the service.

2.7 Pricing

There are also conditions for use of the service that are not directly related to the quality, value or perceived value of the service. People make a trade-off when they decide on how they are going to spend their money.

Various kinds of revenue models exist for Internet services and many of them can also be applied to mobile information services. The usage of the terminology revenue model and business model is sometimes not clearly distinguished. In our view the revenue model concerns the way the financial transactions are arranged and the business model is a broader concept. Rappa (http://digitalenterprise.org/models/models.html) provides an overview of what he calls business models and revenue models can be directly derived from these Possible revenue models to create income by offering mobile information services are the following.

Advertising; The web advertising model is an extension of the traditional media broadcast model. The website publisher provides content (usually, but not necessarily, for free) and services (like email) mixed with advertising messages in the form of banner ads. The advertising model works best when the volume of viewer traffic is large or highly specialised. Advertisement on the mobile internet is even more difficult than on the internet due to the small screens. Clicking to links via a pay-for-performance model is more opportune. This is the so-called *affiliate* model which, in contrast to a generalised portal, seeks to drive a high volume of traffic to one site. It does this by offering financial incentives (in the form of a percentage of revenue) to affiliated partner sites. The affiliate model is inherently well-suited to the internet and the mobile internet.

Subscription; Users are charged a periodic – daily, monthly or annual – fee to subscribe to a service. It is not uncommon for sites to combine free content with "premium" (i.e., subscriber-only or member-only) content. Subscription fees are incurred irrespective of actual usage rates. There are, for example, many mobile services on portals (e.g. i-mode portal) on which one might take a subscription, like the news, sport news, weather, and traffic information. Subscription and advertising models are frequently combined.

Utility; The utility or "on-demand" model is based on metering usage, or a "pay as you go" approach. Unlike subscriber services, metered services are based on actual usage rates. Traditionally, metering has been used for essential services (e.g., electricity water, long-distance telephone services). For mobile services the amount of data transferred is often charged by the actual usage.

Merchant; this is the revenue model for wholesalers and retailers of goods and services on the Internet. Sales may be made based on list prices or through auction. It is e.g. possible to buy a ringtone from a ringtone merchant. If this ringtone is bought directly from the content creator it is the *manufacturer* or "direct model",

Community; The viability of the community model is based on user loyalty. Users have a high investment in both time and emotion. Revenue can be based on the sale of ancillary products and services or voluntary contributions; or revenue may be tied to contextual advertising and subscriptions for premium services. The Internet is inherently suited to community business models and today this is one of the more fertile areas of development, as seen in rise of social networking. Mobile phones facilitate communities since people can access the community anytime anyplace. Even the combination of a real-life community and virtual community can be achieved by mobile services.

The revenue models offered by service providers dictate how the customers pay for the services. Pricing is a reflection of user value. Pricing is not only about 'how much' it costs; it is also about the price structure. Price levels for mobile services are always debatable and will be decided in the market. The pricing model (structure) should be part of the design process. Decisions have to be made regarding paying per minute or per data package and paying per subscription or per use. The option to pay per minute or data is also dependent on the chosen technology. The call to provide price transparency for customers has not yet been met.

2.8 Design example: Buddy on Demand service

This design example has been developed for the travel support domain. The service system in the design example is aimed at foreign truck drivers who drive to and from the port of Rotterdam, and offers a mobile community to them. The truck drivers might be interested in meeting their fellow countrymen during their breaks in restaurants. The service sets up meetings between friendly truck drivers, while taking their preferences and driving schedule into account. The design example focuses on the service concept of the service system, and visualises the user interaction with the system by a storyline and a storyboard.

2.8.1 Storyline

Imagine a user of the service, called Jurek. Jurek is a Polish truck driver, who drives between the port of Rotterdam and his home country. Before Jurek can use the service, he has to create his personal profile on the website of the service. He can reach the website over the fixed broadband Internet or by using his mobile handset. His profile consists of his personal data and a list of his contacts. Jurek can add truck drivers that he knows to the list at any time. Jurek can be added to the contact list of other drivers after he has registered for the service.

Now that Jurek has added his profile, he would like to use the service for his ride to Rotterdam this week. He leaves on Monday morning and is due to arrive in Rotterdam on Wednesday at 20.00. Jurek has entered this data on the website, so the service system knows his schedule. The data can be edited during the ride by Jurek, for example, if the time that he is expected to arrive at the port changes. He can do this while on the road, by using his mobile device.

When leaving Poland on Monday morning, Jurek switches on his mobile device. The mobile device determines the location of Jurek every few minutes by using the GPS device in the truck. The software on the handset also detects when Jurek is taking a break. When it appears that he has not had a break for e.g. two hours, the device will start to transmit coordinates to the service system. Of course, this time interval can be altered by the driver. Jurek can also register himself that he wants to have a break, after which the sending of coordinates to the service system is triggered.

At 18.00, Jurek has been driving for three hours without stopping, so the handset is transmitting coordinates to the service system. His friend Roman is driving fifteen kilometres ahead on the same road as Jurek. Roman is also subscribed to the service, and is a member of Jurek's buddy list. Roman also has not had a break for some time.

At this point, the system makes a match between Roman, Jurek and a place where they can rest next to the road. Then, it sends a meeting proposal to the handsets of both drivers, and asks whether they would like to meet each other in a restaurant twenty kilometres ahead. If Roman and Jurek both accept this proposal, they will meet in this restaurant.

2.8.2 Storyboard

The storyboard in this section will show how the user interacts with the service system.

Step 1: Before the truck driver can use the service he has to register. In order to do so, he visits the website of the service provider by using a personal computer or a handset that is connected to the Internet. On the website the user fills in a form.

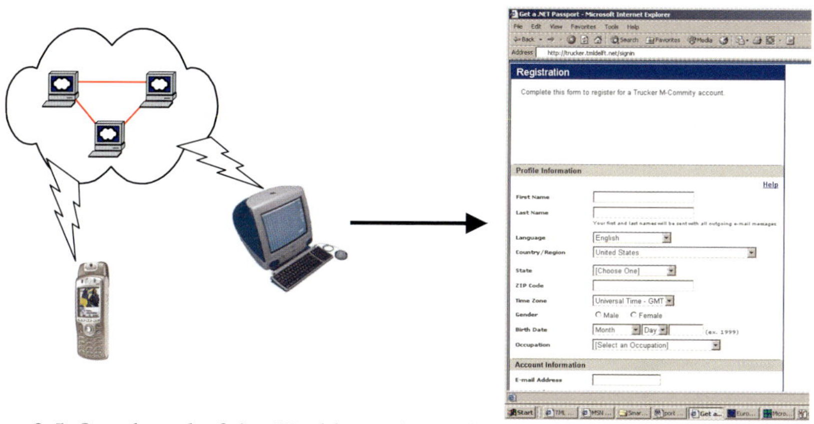

Figure 2-5: Storyboard of the "Buddy on demand" service: Registering on the website

Step 2: Before the driver starts his journey, he has to edit his personal profile. The driver has to insert his driving schedule, such as the destination and estimated time of arrival. He can do this by using his personal computer or his mobile device.

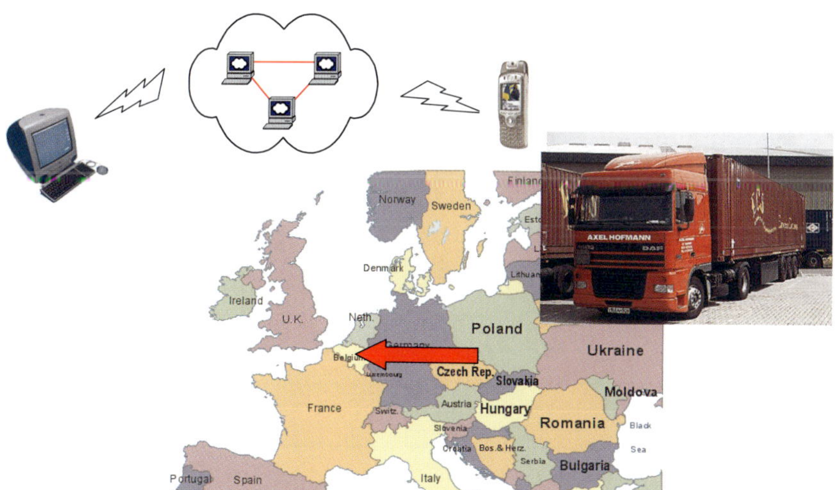

Figure 2-6: Storyboard of the "Buddy on demand" service: Entering schedule data

Step 3: During the ride, the service system will come up with personalised meeting proposals. When both drivers accept the proposal the meeting can take place. In this step, voice recognition is used to ensure a safe use of the service.

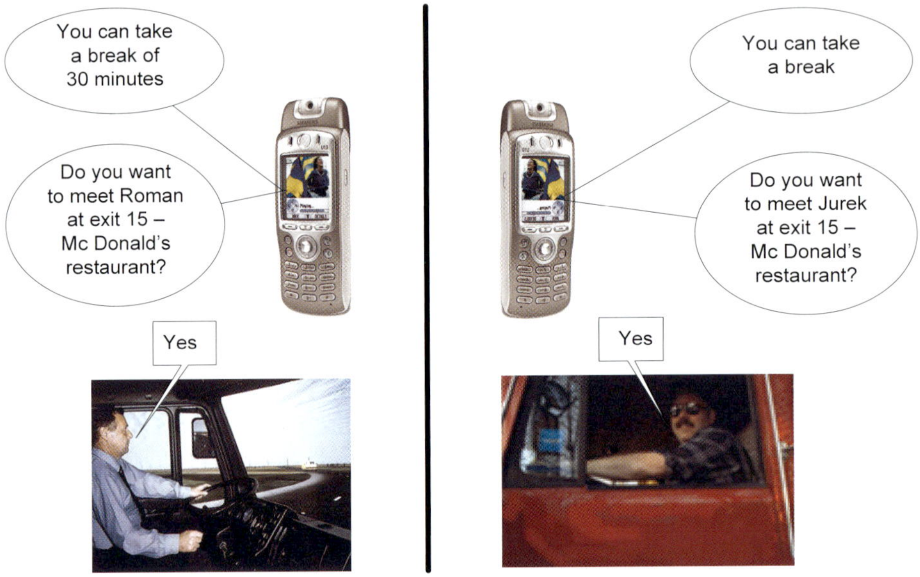

Figure 2-7 Storyboard of the "Buddy on demand" service: Receiving a meeting proposal

2.9 Design example: *SWIFT* Band

This design example has been developed for the information and entertainment domain. The service system in the design example is aimed visitors of a concert and offers wristband with information and transaction possibilities. The design example focuses on the service concept of the service system, and explains the service by a storyboard and a storyline.

2.9.1 Storyboard

The storyboard in this section shows how the user interacts with the service system in a visual way. Each of the 6 pictures are explained in the next section that describes the storyline.

1.
TICKETS
Receive your wristband with your concert ticket
2.
Quicken your ticket validation and access
3.
Make wardrobe tickets obsolete
4.
Wait shorter and pay more quickly and safely
5.
Comfortable to wear and replaceable if lost
6.
Add funds and check detailed overview online

2.9.2 Storyline

The storyline that belongs to the numbered storyboard items above can be found here.

1. Imagine that with the purchase for a ticket of a concert, you received a wristband with an integrated Radio Frequency Identification (RFID) chip. If you log-in online, you can log into a system where you can raise the credit of your account to spend during the concert. Arriving at the event, you could just walk through a gate where you move your wristband past a sensor, where the system recognises the SWIFTband and lets you through. Security does not need to check your ticket for authenticity. After passing security, other visitors to the concert will have to get in line again for the token dispenser. Since you already raised your credit online at home, you can just walk straight to the bar to get your drinks. To encourage people to raise their credit in advance a special combo with beers and chips has been offered. The usage of the service is fairly simple. The most difficult part is the online registration. Our target group consists of mostly young adults who are familiar with internet portals. In addition, the service system can be used without using this registration because the SWIFTband functions as a ticket and is also distributed through the current channels.
2. As RFID chips are difficult to copy, time consuming scanning of tickets for authenticity as happens today will not be necessary.
3. The wardrobe is a normal selling point with a product called "wardrobe" against a price that can be zero or a small amount. Numbered tickets aren't necessary because the number can be assigned to the visitor ID. This requires extra functionality from the transaction application.
4. When a customer orders some drinks, the bartender enters the order on a touch screen, which should be clearly visible for the customer. When the customer agrees with the order on the screen he moves his wristband along the scanner and the screen indicates that the bill has been paid. At the same time the bartender puts the drinks on the counter and the customer takes his drinks. This means the bartender has to enter the order instead of taking, counting and putting away the tokens. He has to reach for the counter only once and he will never have to wait for customers to reach for their tokens. Customers can even enter their order themselves if the touch screen is placed at their side of the counter. This saves the bartender even more time.
5. Because the RFID chips are small and robust they can be easily incorporated into a festival type of wristband. If the wristband gets lost it can be blocked in order to prevent others paying with it. The credits can be transferred to a new RFID tagged wristband.
6. To reduce the queue at the token registry at the start of the event, people can raise their credit by transferring money to a SWIFTband bank account quot-

ing their visitor ID. Using these IDs, visitors can create an account and log in to the SWIFTband web server. There they can check whether their credits have been raised. After the event they can get the statistics of how much they have ordered where and when.

2.10 Conclusions

The service concept is the mirror of what the user demands. In this chapter we looked into mobile information services: what are these; how can they be characterised; how to provide value and increase the chance that users will accept those services. The acceptances also depend on the context of the user. New technology should enable easy-to-use mobile services according to the context of a user, but designing these services is not an easy task. The contextual nature of mobile service needs makes it hard to reveal user requirements. Another factor that makes it difficult to sketch service concepts is the lack of knowledge as to which value propositions will be rewarded by which customers. Surveys of user requirements for mobile services too often support optimism in that users are favourable to a service or feature in abstract but it often appears that it does not fit into their situational needs in practice. Therefore the context of the user should be taken as the starting point in the development of new services. We shall elaborate on this in the design method chapters 5 and 6.

3 Technical architecture

Organizations previous operating separately need to cooperate to offer services to their customers. Often no single organization or department can provide all the required elements and even within one organization the departments involved will have very different architectures. Each organization provides one or more of the building blocks to create the mobile service. Consequently, many organizations need to cooperate in a inter-organisational network and each organization is involved in providing the service. The creation of such a network is complicated as all organizations have different and heterogeneous types of systems that cannot be easily connected to each other. Moreover, collaboration puts high requirements on the technical systems. The systems should be scalable, robust and reliable, as the performance is dependent on the weakest element and adaptive to be able to change to changing circumstance.

In this chapter we will provide an overview of the foundations of the technical architecture. This chapter is structured as follows. In the next section we will discuss the playing field and evolution towards today's situation. This will explain why we are confronted with the current technical architecture. In the section thereafter we discuss the use of service-oriented architectures for the design of service systems, and the use of layered components to construct service systems. After that, we go in more details by discussing 3-tier architectures. Next we discuss network infrastructures and finally we present potential future developments.

3.1 The playing field

Services provided to users consist often of a composition of services which are provided by different organizations over a network infrastructure. The network infrastructure provides the facilities enabling communication and can include basic services as authentication of users and billing services for the use of the network. Fortunately, the network infrastructures is often standardized and shared among parties, which enables connectivity. New end-user services can be created using a pool of existing and new services operated over the network infrastructure.

Figure 3-1 schematically depicts the use of services provided by different organizations. Each organization can provide a number of services and the user is presented a composite. In this view organizations become service providers and the user is a service requester. Between and even within one organization there might be services that rely on or make use of functions and information provided by services provided by external or internal service providers. So one department might act in their role as service provider to the user and at the same time be a service requester by making use of other services. This creates a complex pallet of dependencies and relationships.

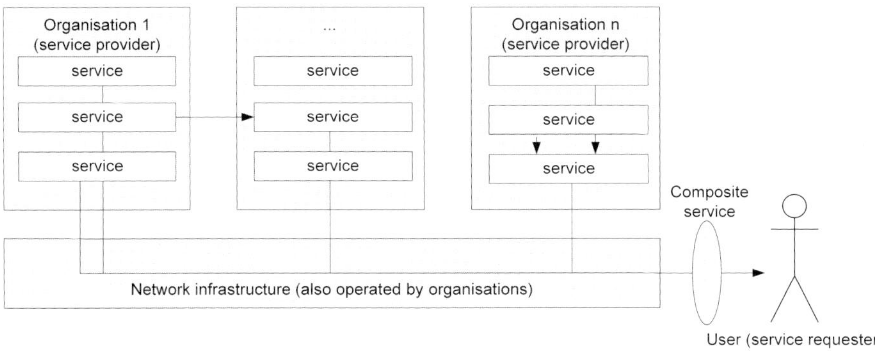

Figure 3-1. The creation of a user service by composing services from various organizations in an inter-organisational network

New services are more and more created in real-time for one unique customer dependent on the user actions and needs. In this way services are produced and consumed simultaneously. The composite service is often created by invoking a number of services provided by the organizations acting as service providers. In this respect, *composite services* are a series of software invocations using a predefined request resulting in a predefined response. A service can be a composite, made up of a number of other services provided by one or more service providers. Each service in the composite service might be a composite service in itself Services can be executed in parallel, at the same time, or sequentially, i.e. one after each other. Furthermore, the answer to a service invocation can be immediate or delayed. In this sense the term services is often used to describe a variety of electronic interactions ranging from basic services, such as the delivery of news, traffic information, games, processing a payment and the take out of an insurance policy, to more complex services, such as the delivery of context-aware, personalized services.

How did we end up with this mess?

In the early days of computing, data was often processed on centralised mainframes. Processes and data existed within homogenous environments and there was hardly any communication between systems. Software vendors had their own, non-standardised platforms and with the development of technology, all kinds of new platforms were created. Over time, smaller and more opens platforms, such as Unix and MS-Windows platforms, which could run all kinds of applications, challenged traditional mainframes. Nowadays, millions of personal computers can be found on millions of desktops within organisations and often no two computers are exactly alike. Application development also changes by adopting object-oriented and component-oriented development paradigms. In addition, the Internet became the preferred vehicle for communications and the number of network connections has been growing with an exponential speed.

In the rush to adopt these new systems, most companies applied minimal architectural foresight for the selection of platforms and application. Many of the IT decisions were based on the perception of the current technology market. The installation of these systems has been a textbook example of "management-by-magazine" (Linthicum, 2000). Rather than making business-driven decisions or using structured decision-making techniques, the decisions were often technology-driven. The decision-making process lacked the perspective of integrating systems as part of the architecture of the already installed base. The situation was even worsened as IT-decisions were made more and more at the lower levels within the organisation hierarchy. Each department was free to select its own technology and applications. As a result, within (and between) organisation(s) many stovepiped applications were created that addressed and solved problems within departments. The systems were built without giving considerations to the requirements for information sharing. As a result, there were often no application programming interfaces (API), open architectures, standard data formats and standard terms and definitions. The data stored in various systems might be overlapping, e.g. name and address information in multiple systems, stored in different data-formats, e.g. "Marijn Janssen", "M. Janssen", "MFW Janssen", "M.F.W.H.A. Janssen", and using different formats for storage, e.g. text, binary formats, XML, and so on. Moreover the quality of the data might be different and data was often not up-to-date. For example, changes in address information are often not communicated to all parts of the organisation that have systems storing address information. Even within one company address information within systems might not be up-to-date. The end result is a chaotic environment consisting of heterogeneous software, hardware and other systems. These systems are often addressed as *legacy system*, to show that these systems are based on outdated technology, which can often not easily be accessed.

The technical architecture for new services need to be integrated with the legacy systems found in various departments and organisations to enable the reuse of data and functionality. The systems architecture has to deal with a heterogeneous information systems landscape and cannot be based on a greenfield approach. Therefore, when designing new service systems, it is very important to analyze already existing systems and to see if and how an interface to these systems can be made and what the cost of that interface will be.

With the rise of mobile technologies, various types of devices appeared, running different operating systems, and having various screen sizes. The constraints like limited user interface, small form factor, low processing power and low bandwidth might limit the services that have to be developed for these devices. To enable presentation and connection to all types of systems, a certain degree of standardisation is necessary. Furthermore, the installed base of handsets already in the market may hamper the use of services that require the newest technology.

3.2 Service-oriented architectures

Traditionally, the emphasis has been on developing systems for use within one organisation or department without considering the need for interoperable systems and processes. This has resulted in a huge interoperability problem as all organisations have their own heterogeneous types of systems. With the arrival of the service-oriented architecture paradigm and the standardization of web services protocols, the technological threshold for creating interoperable systems has been lowered. Software systems are viewed as providing services to *service requesters*, which could be other systems or users. Other systems can remotely request information or invoke some kind of functionality from the service. As such, software systems act as *service providers* and can be considered as consisting of a set of services. Organisations can describe and publish their service descriptions in a directory, and others who want to make use of these services can search the directory, select the right services, and subsequently invoke the services. In the e-services paradigm, the world is viewed as a collection of service providers, requesters and brokers (yellow pages) as shown in the figure below.

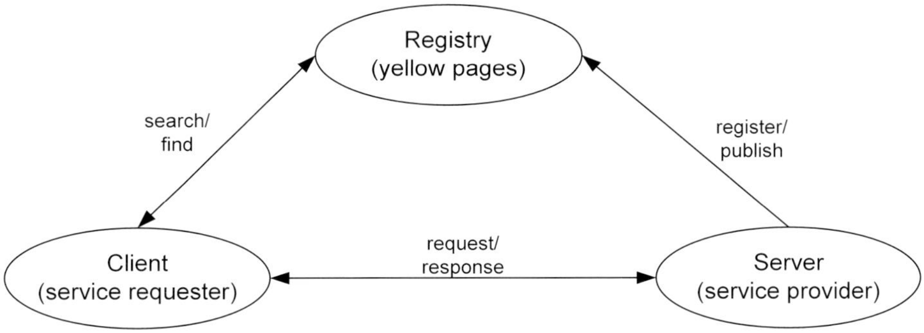

Figure 3-2. The basic idea of a service-oriented architecture

Service-Oriented Architectures (SOAs) define how computing entities can interact in such a way that one entity can perform a unit of work on behalf of another entity. An SOA is essentially an architecture that describes the communication pattern between services communicating with each other, with functions and with their operations (Papazoglou & Georgakopoulos, 2003). In the perspective of SOA, a service is a function that is well defined, self-contained, and does not depend on the context or state of other services. SOAs do not have any specific technology or a description language. There are, however, a number of technologies that support the so-called web services protocol stack, which are often viewed as the de facto standards for SOA.

A service-oriented architecture utilises a firm's existing assets and applications as services to integrate business processes and supporting business needs. A well-implemented SOA can provide adaptiveness and agility for a business by aligning IT infrastructure with business needs by taking advantage of its existing assets and modularising and coordinating systems, components, and other resources as services to support new and existing business processes. *Agility* concerns the use of interoperability standards to minimize the time and costs of deploying new applications. *Adaptability* is the ability to deal with new environmental conditions. Higher levels of adaptability are able to deal with more complex changes in the environment.

The SOA paradigm facilitates the creation of modular, accessible, self-describing, implementation-independent, interoperable and reusable components that can be published and remotely invoked (Fremantle, Weerawarana, & Khalaf, 2002). Generally, a service communicates using well-defined, standardised interfaces, independently of other components. The manageability increases, because large modular services are constructed from smaller ones. In essence, a complex problem is split up into smaller problems, which can be solved independently. Openness and flexibility is created as new services can be added and removed from the architecture and each single component can be replaced by another component without affecting the others. Service-oriented, modular architectures can leverage investments in legacy systems running the key business-critical applications (Arsanjani, 2002). Modularity has several major advantages:

1. it increases the range of manageable complexity during development and deployment,
2. allows different parts of a large system to be worked on concurrently,
3. enables reuse, and
4. accommodates uncertainty.

In essence there are two main strategies for developing information systems, "frameworks" and "best-of-breed" (Lee, Siau, & Hong, 2003). A best-of-breed strategy is based on the concept of modularity.

Frameworks are often viewed as synonymous with Enterprise Resource Planning (ERP) systems, however, there are many other types of frameworks. An ERP system is an enterprise-wide software solution designed to streamline the data flow between different functions in an organisation. In essence an ERP is a piece of multi-module application software that assists a business in managing the important parts of the business, including financial reporting, planning and control, purchasing, customer relationship management, and collecting management information. Often ERP packages contain a whole range of adapters to connect to other systems. An ERP provides all business functionality in one package that can be configured to match a business. The ERP implementation is technology push oriented, and an ERP system forces the business to adapt the business processes to the functionality offered by the software. ERP systems are often bought by one organization and not by a complete interorganisational network. If the other network members use the same ERP system, the creation of interoperability is relatively easy and straightforward. If not it might be cumbersome and very difficult.

Best-of-breed approaches focus on assembling a system from modular components that are 'best' on some aspect, e.g. cost, performance, functionality etc. A system is built by integrating the individual components. The premise of best-of-breed is that it allows a business-oriented approach to map software on business processes rather than technology-driven business process reengineering. The current trend is to make use of service-oriented architecture in the design of frameworks. In this way frameworks also modular as organizations can only use those modules that match their needs.

In this chapter we focus on 'best-of-breed' applications. Often these modular components are connected using service-oriented architectures. In such a strategy managers focus on decision-making around the selection of the best-of-breed components. Architects are involved in this decision-making, and should have to possibilities and impossibilities in mind of integration best-of-breed applications in to total landscape.

3.3 Technologies for service-oriented architectures

There are many technologies that can be used to create a service-oriented architecture. The most notable that is used more and more are web services technology. Web services provide machine-to-machine communication to enable interoperability between (distributed) applications. Web services provide a way for sharing information and functionality using standard protocols crossing departmental and organisational borders. In this way existing information and functionality can be used when building new service systems. Although web services enable technological interoperability, there is still a need to standardise the data formats of the interfaces.

The paradigm of web-services is often associated with modularity (Janssen & Wagenaar, 2003). With web-services, software functionality and data can be "packaged" and accessed as a web service. Wrappers and adapters often make legacy systems accessible by providing an interface based on standard protocols. Many different types of modules can be found, described, discovered, and integrated using web-services technology. Service-oriented paradigms are becoming more important in today's design of information systems. Service-oriented business integration enables the on-demand composition of new business processes using already existing services possibly provided by other parties. In service-oriented applications, services are configured to meet a specific set of requirements at a certain point in time, to be executed and then disengaged. Services only 'exist' during execution, and modules provide services.

The World Wide Web Consortium (W3C) defines a web services as "a software system designed to support interoperable machine-to machine interaction over a network" (www.w3c.org). The goal of web-services is to abstract business logic from implementation. In this way it should provide a way for communication between applications that is platform, vendor and application independent. In fact, web services perform encapsulated business functions, so that there is no need to bother with implementation details. The term web-services has already been overloaded, and it could be stated that,

- *Web-services* (plural) is a blanket term referring to all technologies that enable interoperability.
- *Web-service* (singular) is a term referring to a collection of application operations accessible over a network using web-services technologies.

Web services technology also incorporates a technology for orchestrating the sequence of web-service invocations into complete business processes. It can be used to coordinate processes residing in various heterogeneous environment and can be used for creating workflows (Verbraeck, Tewoldeberhan, & Janssen, 2006). In this sense the paradigms of workflow management and service-oriented become closer together.

Mobile Web Services (MWS) are the application of web services technology to the mobile environment. The standard web services technology can be directly applied to applications on mobile devices. As web services concern machine-to-machine interactions, the use of mobile web services requires that an application is running on the mobile device. The application on the mobile device can find web services and then invoke them. In combination with the "anytime and anywhere" availability of mobile devices and location information, web services can be used to design new innovative mobile services. A service is often composed of a number of standard web services.

3.4 Three-tier architectures

Mobile devices communicate with services on servers using pre-defined protocols. Mobile networks themselves provide additional services such as user location and user identification that are used by mobile applications. Current architectures often consist of multiple tiers aimed at creating a modular architecture to increase the flexibility, usability and scalability in comparison to traditional systems. In a three tier architecture, a middle tier containing the business logic is added between the client tier that provides the user interface to the user, and the data tier that contains the relevant data for the service. Figure 3-3 shows an example of a three tier architecture.

Although the three-tier model suggests a relatively simple architectural design, the reality can be much more complex. Different servers can be in place in each tier. Software systems that are used to implement the functionality for one tier, can also contains functions for other tiers. An example is a database application that contains application logic and presentation modules. Separating the functions into tiers using the available applications is not an easy task.

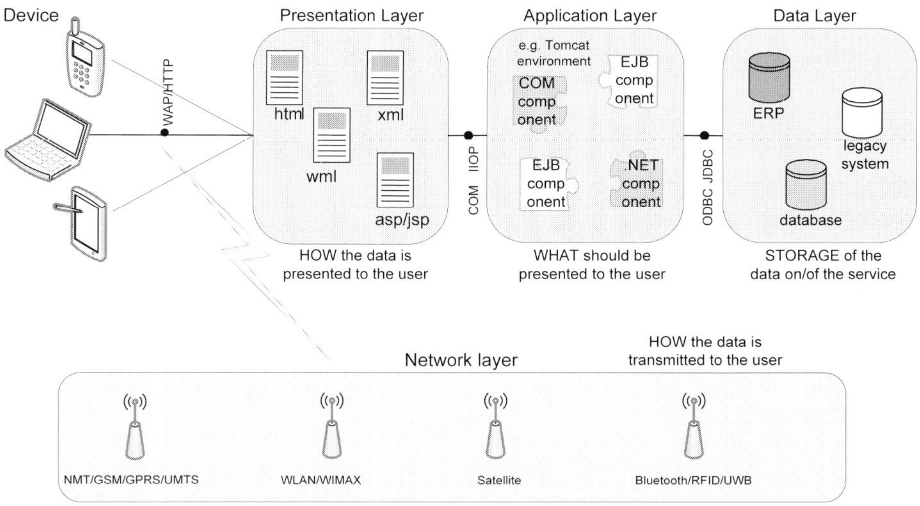

Figure 3-3 Three-tier technical architecture overview

Although 2-tier, 4-tier and n-tiers architectures exist, typical current architectures use a three-tier model, as shown in Figure 3-3. Tiers are also called "layers", because one can think of the presentation layer sitting on top of the application layer, which uses and updates the information in the data layer.

- *Presentation layer*: The presentation tier can either generate the data in the right format to be transmitted to an end-user device, for instance to display information to the end-user, or generate data to communicate to another application, either in the end-user device, or on a server. For presentation in mobile devices protocols like WML and WMLScript are often used. WML stands for *Wireless Mark-up Language*. It is a mark-up language that like HTML has been based on XML. WML uses WMLScript to run simple code on the client. WMLScript is a light JavaScript language. In contrast to JavaScript, WML scripts are not embedded in the WML pages.
- *Business logic layer*: The middle tier typically includes application logic and business rules, and it can contain messaging and transaction processing technology. The application layer often contains the business logic for managing business processes. A so-called *server* often makes up the application layer. The term server is not used to refer to a piece of hardware, but to a software programme that runs

on a computer to execute componentised pieces of software. An example of an open source Web server programme that is often used is Apache Tomcat.

- *Data layer*: This tier stores information, which could be pictures, data, geographical information and so on. Technically, the data layer includes relational and object databases, information systems, and other data sources.

A key decision in a multi-tier design is how different tasks involved in delivering a working application to the users are distributed over multiple tiers.

Mobile devices tend to have a more constrained environment (less bandwidth, smaller screen, less power and less processing speeds) than desktop computers, therefore a separate suite of protocols has been developed for mobile applications. For enabling mobile devices to communicate with remote applications often the Wireless Application Protocol (WAP) or i-mode protocols are used. The WAP forum has aligned its standards closely with the Internet and Web standards for communication protocols and presentation languages. A closely related architecture is the i-mode architecture developed by NTT DoCoMo in Japan. Their standards are aligned even more closely with web technology. Both technologies have converged towards a common Internet-based architecture which can be used on top of network communication protocols like GSM, GPRS or UMTS.

The WAP architecture defines an end-to-end application protocol. The application protocol is a communication protocol embedded in each WAP enabled mobile device. The server side (WAP gateway) implements the other end of the protocol and routes the request from the web enabled mobile device (the client) to a web server. A WAP 2.0 or i-mode gateway service can connect to a GPSR/UMTS gateway support node which acts as the interface the between mobile network and the Internet. WAP services are hosted on web servers that are often connected to the data layer to obtain user information, product information, and other content.

Nowadays the mobile web and the Internet are more and more interwoven. Mobile devices are running operating system and browser similar to the browser in a conventional computer.

3.5 Designing new services

The services and network infrastructures are decoupled and can be combined to create a service composition as discussed at the beginning of this chapter. The complexity of an interorganisational network is reduced by decomposing them in separate, well-defined services. These services need to be integrated in business processes or need to be composed to provide a service to the end-user.

Many organisations are involved in providing the end-user service. Often information needs to be collected and distributed to various organisations to enable a customised response. The content can be provided by making internal information available, and by linking the service system to other information sources. Rather than linking directly to external databases, it becomes common to link to external web services instead and in this way separating the interface and the implementation(s). This enables the replacement of one implementation by another implementation without affecting the other parts.

Services can take care of the non-availability of the data or databases, for instance by implementing caching mechanisms or by providing mirror services. One can think in terms of *delegation* when linking a service to other services. These other services have a delegated responsibility for providing the right information with an agreed quality, and within a time period that is aligned with the user needs and the user processes. This requires a good technical architecture, preferably with standard protocols, and a good organisational network with agreements between the different parties. The challenge is to link the main service and the delegated services in such a way that the user of the main service does not see where the main service stops and the delegated services start. The information has to be truly integrated, and often it is necessary to make a conversion of the externally acquired data to fit the other data that is offered in the service.

3.5.1 Customisation and identification

In addition to the content, the service processes have to be designed as well. The organisational processes determine how the service will be offered: Is the user in control of the sequence of activities? Or is the system in control and is the sequence of activities fixed by the service system? Many systems are a mix of these two extremes, and especially in such a semi-flexible situation, it has to be carefully determined where flexibility can be offered and where the system is in control.

In order to make personalisation work, user *identification* is needed. This might be as simple as a user code / password check, or as advanced as for instance a biometric scan. Often a single sign-on is used to enable the trusted partners in the interorganisational network to exchange user identities. Organisations are offered the ability to authenticate users in order to grant them access to the services provided by these organisations.

Mobile devices are close the users. To avoid that users need to enter information on the flexible parts every time, *personalisation* is often applied as a strategy. The service system stores characteristics and preferences of the user in a personal profile. Personalisation can be implemented and invoked as a web service that governs the delivery of the main service to the user. Settings about the interaction (colours, font size, location of sub-windows, etc.) can be stored, as well as settings about information presentation (e.g., units and language to use), and the sequence of activities (when tasks A and B are independent of each other but must be completed before task C starts, some users prefer A-B-C, and others B-A-C).

The most advanced processing schemes for service delivery also take the user context into account, such as the time of day, the user's location (so-called location-based services or LBS), or the status of the information that is important for the user. An example of a status dependent service is a service for emergency workers where the type of information delivery is very different in a crisis situation as compared to a non-crisis situation. During a crisis, less important information such as incoming e-mails or private calls may be suppressed, and only the information that is relevant for the crisis response can be allowed to pass.

3.5.2 Integrating legacy systems

Often the architecture has to deal with a heterogeneous information systems landscape. In cases where existing information systems, which could be legacy systems, are present in the organisation, these systems might have to be adapted or replaced in order to make the service system work. Redesign of information systems to satisfy the needs of a service-oriented architecture is not easy – often the older systems have been designed with a focus on the data that has to be retrieved and on users who are present from 9 to 5 in the office. In the new setting however, the user is in control, and he/she is very flexible in terms of sequence of activities and times when to use the service. When the existing information systems cannot be replaced to make them more service-oriented, they have to be 'wrapped' to make them appear service-oriented on the outside. The wrapper ensures that the system can be approached as a web service. The activities to adapt the interfaces of existing systems to make them more compatible with new architectures can be an expensive and time-consuming process. Not only is there a mismatch between the old technologies available in the existing system and the new technologies that the services require, often it turns out that the demands that the new services pose on the system are so high that the system cannot function anymore to fulfil its old tasks.

3.6 Mobile network infrastructure

In the past decade a move from traditional circuit-switched networks to packet-switched programmable networks that integrate both voice and packaged services has taken place (Bi, Zysman, & Menkes, 2001). The developments in mobile infrastructure technology are enabling the deployment of new mobile services. The new technologies support faster transmission speed and therefore multimedia data can be transmitted over wireless networks.

The role of security is essential in all parts of the future network architecture. Security should be taken into account in the operation of the device, the radio access network, the core network and the service platform part, and in all layers including hardware, operating systems, protocol stacks and applications, to address the relevant constraints and security threats (virus, DoS, spam, etc.). The great challenge for the security frameworks is to maintain simplicity and efficiency without leading to heavy implementation solutions (Tafazolli et al., 2005).

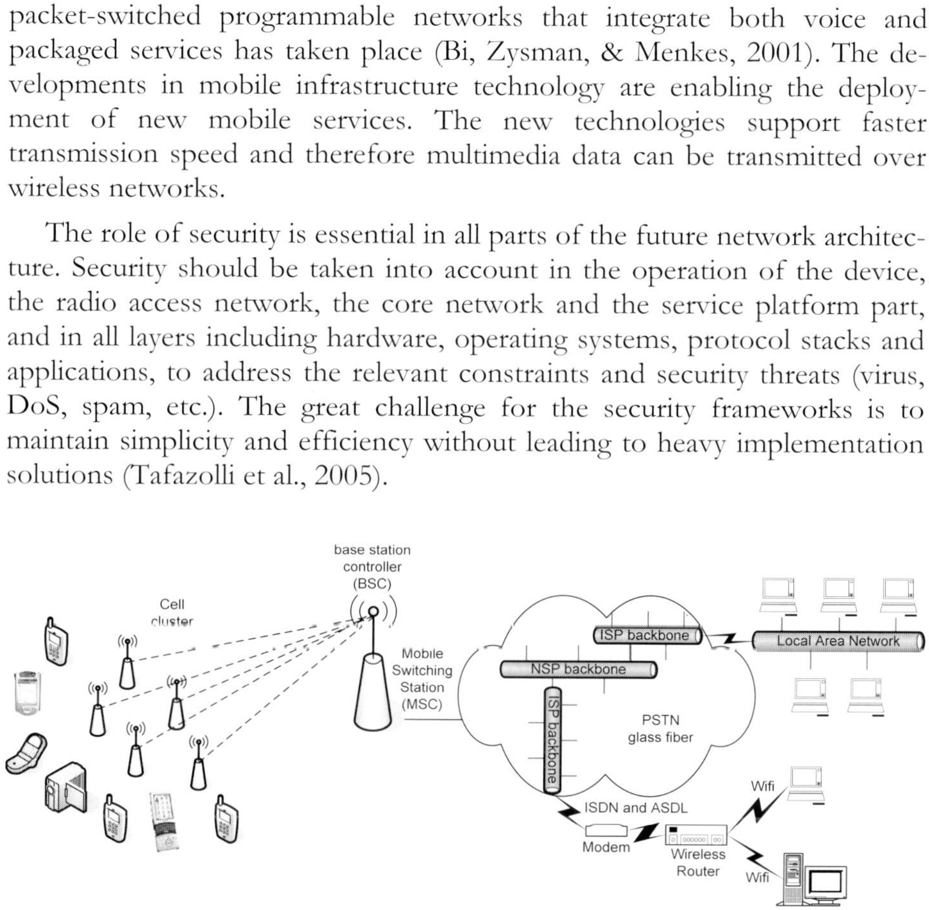

Figure 3-4 Mobile network example

Figure 3-4 illustrates the complexity of the infrastructure of a mobile network. Internet services can be accessed using mobile terminals such as mobile phones or PDA devices through wireless connections (left-hand side of the figure). The backbones are run by the Network Service Providers (NSPs), which include the major telecommunication companies. The delivery sub-networks are provided by the local and regional Internet Service Providers (ISPs). The ISPs exchange data with the NSPs at the Network

Access Points (NAPs). The Public Switched Telephone Network (PSTN) refers to the international telephone system based on copper wires often carrying voice data. This is in contrast to newer telephone networks based on digital technologies, such as ISDN and ADSL.

Mobile technology uses signals that are transmitted through air or space and incorporates a variety of technologies varying in reach, scale and complexity. Current technologies improve the multimedia use of the mobile device, and nowadays all kind of devices use mobile technologies, including PDAs (Personal Digital Assistants), cellular phones, mice, headsets, digital cameras, printers, laptops, etc. Standardisation is often mentioned as the key to technology success.

3.6.1 Communication protocols and technologies

Table 3-1 shows an overview of the type of technologies and services based on the type of network and bandwidth. The geographical scope is used for defining the type of network. Definitions of broadband are subject to change over time, as faster networks speed becomes available. Broadband can be defined as a permanent high-speed connection in excess of 128 kilobits per second. A Personal Area Network (PAN) is a communication network enabling wearable devices to communicate with each other. An example is a RFID chip embedded in products in a shop. A Local Area Network (LAN) supplies networking capability to a group of computers within a short range, such as within a house, office or school. A Wide Area Network (WAN) is a geographically dispersed telecommunications network. The main technologies are positioned in this table. This is a simplified view showing the technologies and the geographical reach of these protocols.

People choose systems in which they look for information and communication depending on the context and circumstances. Behind their desk in the office they use a PC connected through a broadband connection, at the airport they use their laptop connected through WLAN and out in the field they use their mobile phone connected through a network technology like GSM. In all these settings they might search for the same information, for example they look for contact information in a directory service like the yellow pages.

Table 3-1 Mobile technologies

		Personal area networks	**Local area networks**	**Wide area networks**
Broadband	Example		High-speed interconnection of video and audio equipment	Video conferencing, webcams, collaboration tools
	Technology		Ultra Wide Band (UWB)	UMTS, Wi-Max 802.16
Narrowband	Example	Authentication, wireless monitoring, alarms, earphones	Internet access, file sharing, route information	Voice, pictures sharing, payment
	Technology	RFID, Bluetooth	Wifi, IEEE 802.11x	GSM, GPRS, satellite

A current development is to combine technologies with each other to offer mobile access whatever the technology and access mode is. This can provide significant cost-saving for users. Different network providers might use different technologies. Moreover, when moving across borders one might want to switch to a different service provider. *Roaming* refers to the switching from one network provider to another network provider. This means that the network providers have to have some kind of agreement with each other, as the second provider has no direct pre-existing financial or service agreement with the particular subscriber. Agreements can be based on barters, but often all kind of e-services are used to monitor the network use, to process the usage information, to bill each other and to settle the bill.

Personal Area Network technologies

Radio Frequency IDentification (RFID) refers to the technology that uses devices attached to objects that transmit data to an RFID receiver. These devices can be small or large, i.e. in a small one in books or clothes or a larger one on ocean containers. RFID has advantages over bar codes. It has the ability to hold more data, the ability to change the stored data as processing occurs, it does not require line-of-sight to transfer data, and it is very effective in harsh environments where bar code labels do not work. RFID can improve the efficiency and quality of services. RFID opens the way to better care for people in their own homes; more effective management of transport, warehousing and supplies, and simpler systems for accessing public transport and building.

Bluetooth is a universal radio interface enabling electronic devices to communicate wirelessly via short-range ad hoc radio connections (Haartsen, 2000). Bluetooth is an open specification and has gained support from leading manufactures. It is suitable for seamless wireless short-range communications of data and voice between devices, in this way enabling equipped electronic devices to communicate with each other. The objective of Bluetooth is a dual one: ensuring a good autonomy for Bluetooth devices while guaranteeing an optimal use of batteries. Bluetooth makes it possible for two devices to communicate over a distance of 10 metres, however, distances can be increased if outlet power is used. Bluetooth device can function in IP (Internet Protocol) packet communication mode or throughput switching mode.

Local Area Network technologies

Wifi (short for "wireless fidelity") or IEEE 802.11x. provides a hotspot coverage and is suitable for high-speed Internet access services for stationary users (http://grouper.ieee.org/groups/802/11/). Therefore, Wifi is very suitable for businesses and private users who want to create a Local Area Network (LAN). A nomadic worker or researcher travelling around the world has to search for hotspots to make an Internet connection. Physically, a wireless access point is used to connect wireless devices to existing wired networks. From a user perspective, there is not much difference between connecting through a Wifi access point or though a network cable plugged into the computer, except for more mobility. This provides you the opportunity to sit outside and enjoy the weather and at the same time connect to the Internet.

Wide Area Network technologies

WiMax (Worldwide Interoperability for Microwave Access) or IEEE 802.16 is a broadband wireless access system using point-to-multipoint infrastructure operating at frequencies between 10 and 66 GHz (http://grouper.ieee.org/groups/802/16/). Next versions support latency applications, e.g. voice and video, and provide broadband connectivity without requiring a direct line of sight between the terminal and the base station. WiMax extends the hotspot coverage of Wifi to Metropolitan Area Networks (MAN). WiMax uses the same technology as Wifi, but has the potential to become a widely used fast communication service like GPRS and UMTS.

Cellular phones transmit their signals to a grid of cellular stations that are linked to the wire-based telephone network (Bi et al., 2001). Coverage is created by a set of overlapping cells that collectively form a cell cluster as shown in Figure 3-4. Each cell uses a base station to send and receive signals to and from mobile devices operating within the geographical reach of each cell. The signals are communicated to a base station controller (BSC) that is connected to a mobile switching centre (MSC) in order to communicate with the PSTN. A cellular phone operates only if cellular stations supporting the various types of transmission protocols are available. If so, cellular phones provide full mobility.

The first generation of cellular technology was analogue based. The second generation is widespread nowadays and is known as Global System for Mobile Communication (GSM), which uses the 900 MHz and 1800 MHz frequencies in Europe, Asia and Australia, and the 1900 MHz frequency in North America and Latin America. GSM is a digital cellular technology that is used predominantly in Europe and Asia. GSM is the world's leading standard in digital wireless communications. General Packet Radio Service (GPRS) is a radio technology for GSM networks that adds packet-switching protocols and is often know as "generation 2.5" of mobile phone technology.

Universal Mobile Telecommunications System (UMTS) is a third generation mobile communications technology (http://www.umtsworld.com) that provides wideband code division multiple access (CDMA) radio technology. The CDMA technology offers higher throughput, real-time services, and end-to-end quality of service (QoS), and delivers pictures, graphics, video communications, and other multimedia information as well as voice and data to mobile wireless subscribers. UMTS offers tele-services (like speech or SMS) and bearer services, which provide the capability for information transfer between access points. It is possible to negotiate and renegotiate the characteristics of a bearer service at session or connection establishment, and during an ongoing session or connection. Both connection-oriented and connectionless services are offered for Point-to-Point and Point-to-Multipoint communication. Both GPRS and UMTS (Universal Mobile Telecommunication System) are evolutions of the GSM networks.

Less known is the use of satellites moving in geostationary orbits that remain above the same part of the earth. Although it is not adopted on a large scale in many western countries, it provides many advantages. It can be used in remote areas, as it does not need a wire channel and does not need earth-bound stations. For example ships, oil platforms, and remote areas make use of satellite communication.

Global Positioning System (GPS) is often associated with location-based service system. UMTS networks can support location service features, to allow innovative location based services to be developed. It will be possible to identify and report the current location of the user's terminal in a standard format (e.g. geographical co-ordinates), and to make the information available to the user, mobile device, network operator, and service provider. Geographic Information Systems (GIS) are important for location systems as they provide the tools to provision and administer base map data such as man made structures (streets, buildings), terrain (mountains, rivers), and important facilities such as gas stations.

The network operators are in important component of the total service system. They provide the connectivity over their network structure and basic services such as user location and user identification that can be leveraged by mobile service applications.

3.7 Future developments

The future is hard to predict and many predictions come not true, however, there are a number of developments that will most likely influence the future. We expect the mobile industry to keep playing a dominant role in the coming years and mobile communication to become the usual mode of operations. Ubiquitous high-speed wireless networks will become available everywhere and mobile hardware can become part of, or even integrated in, everyday's outfit. In their technology roadmap for a wireless 2010 vision, Mihovska et al. (2006) describe several wide and metropolitan-range technologies and networks of which they expect that they will be available by 2010. In addition, they expect that will be an increase in the use of short-range network technologies. These can be used for the control and monitoring of system operations or daily activities. Many products will contain embedded RFID chips that allow monitoring of products when they move through an interorganisational network.

One of the key developments is to be able to switch seamlessly from one type of network to another network. For example at a railway station a connection to the installed Wifi spot can be made, when travelling one of the cellular phone networks can be used. This provides the opportunity to select the lowest cost or the fastest network available. Another trend is the move from narrowband to wideband with a family of standard tailored to a variety of applications needs (Bi et al., 2001).

The network speed likely increases and broadband likely becomes available anytime, anyplace. Mobile devices will have more colourful and easy-to-use interface able to display more data. Integration of GPS technology with wireless networks will continue and enable more location-based applications and emergency services. More advanced privacy and security mechanism might be developed to better ensure the privacy and security of mobile users.

3.7.1 Services become part of the network infrastructure

Many current efforts in the mobile network infrastructure field are aimed at creating seamless connectivity. Devices operated by users can be roaming throughout different service providers' networks that may use different technologies without losing connectivity. In this way advantages from different technology can be used and it will allow universal mobility and a high data rate without users noticing the change in providers.

Another tendency is to include more and more basic services as part of the network infrastructures. In the past the mobile network infrastructures were primarily the cables and the protocols to connect one location with another. Currently, services like authentication and billing have become part of the mobile network infrastructure.

3.7.2 Convergence of web services and semantic web technology

In the world of web services the convergence of web services and semantic web technology takes place resulting in Semantic Web Services (SWS). SWSs are often advocated as the next major generation of the Web, in which services and business communication become more knowledge-based and agent-based. In its simple version *semantic web* means revealing information related to web data, with the goal of making searches on the Internet more precise. For example a mark-up tag on a web page could describe a document's content and its key links to other documents. Dynamic access interfaces are improved by defining the types of parameters of the interface's operations more precisely. On the other hand a completely fresh look at the storage, representation and access of web- based information and knowledge resources is advocated. The champions of this viewpoint aim at "unlimited" meaningful and knowledgeable access to distributed information resources in real time. The software will automatically profile users and assemble and harmonise diverse information. Interfaces to software systems and enterprises are defined using new means to describe their semantic behaviour in order to almost automatically compose dynamic behaviour. As such semantic technologies are a necessary technology for effective discovery, matching and integration of web services in business processes.

3.8 Design example: Finding disaster victims service

The service in this design example is developed for the emergency domain. It is intended for relief workers at the site of a large disaster in a residential area. In this situation, it is crucial for relief workers to know where potential victims of the disaster are. This can become complicated if the regular GSM or UMTS networks go down, because they are overloaded or even destroyed. The service in this design example enables the relief workers to communicate via an emergency network. This network is used to inform them on how to get to the location of potential victims of the disaster.

The design example description starts by briefly introducing the service concept. Next, the technical architecture of the service system is discussed in more depth. Finally, the steps and interactions that are involved in getting the information on the whereabouts of the victims to the relief workers are described.

3.8.1 Service concept

When a residential area is struck by a major disaster, the mobile networks in that area may become overloaded or might even be destroyed. That makes these existing wireless networks unusable for emergency workers. At the same time, it is essential that the relief workers in the area are well informed about the location of potential victims, in order to make the rescue operation as effective as possible.

The service in this design example consists of two main features. The first feature is that it provides an emergency network that can be used by relief workers to communicate with the outside world. The second functionality is that it sends information to the relief workers on how to get to the victims of the disaster.

3.8.2 Technical architecture of emergency network

We will first discuss the technical architecture of the emergency network as such, see Figure 3-5. It must be possible for the relief workers to communicate with each other, and with the outside world. Therefore, the architecture of the emergency network has to consist of two connected parts. These are the communication link to the emergency centre, and the emergency network at the scene of the accident.

Many technical options are possible to establish a connection from the disaster area to the calamity centre. Most options, such as GSM, GPRS and UMTS are on ground level and are therefore at risk of failing in the event of large disasters. The only communication infrastructure that is beyond reach of almost any possible disaster is the satellite network. Therefore, the calamity centre will be connected to the scene of the emergency through a communication satellite. To do so, a mobile station, e.g. a vehicle, will be equipped with a satellite receiver and dispatched to the area of the emergency.

An emergency network is also needed at the scene of the disaster. To do this, the mobile stations that have a satellite receiver will also be equipped with a Wifi access point. The relief workers who carry laptops, palmtops and Voice over IP phones will be able to connect to this access point. In that way, they can communicate with each other via the mobile station.

When combining the satellite link and the Wifi emergency network, the relief workers can also communicate with the emergency centre. First they have to connect to the mobile station using Wifi, and then the mobile station forwards the communication via the satellite link to the emergency centre.

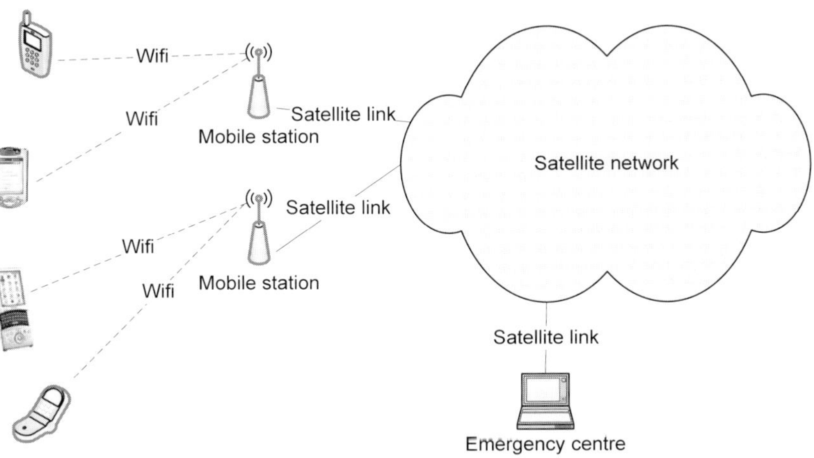

Figure 3-5. Technical architecture of emergency network

3.8.3 Sequence diagram for localising potential victims

The service system will use the emergency network to send directions to the relief workers on how to get to the victims of the disaster fast. Of course, the emergency network could be used in many more ways, such as communication between the relief workers, or exchange of voice and video messages between the relief workers and the emergency centre.

Figure 3-6 visualises the steps between the moment that the relief worker sends a request on where to find the victims and the moment that he/she receives the instructions where to go. First, the relief worker enters a request for the location of nearby victims on the mobile device. The mobile device forwards this request to the emergency centre, by using the emergency network. The emergency network also determines the position of the relief worker. Now, the emergency centre sends a request to the mobile network operators having GSMs near the location of the relief worker. The mobile network operators use the databases in their network to determine the last known location of their subscribers at the location of the relief worker. Then, if still possible, the mobile network operators perform triangulation to get a more precise indication of the current location of the subscribers.

Then, the mobile network operators return the location of their subscribers who are near the relief worker. The emergency centre uses this information to generate route instructions for the relief worker on how to get to these victims fast. The route instructions are sent to the mobile device, via the emergency network, and the relief worker reads them from the screen.

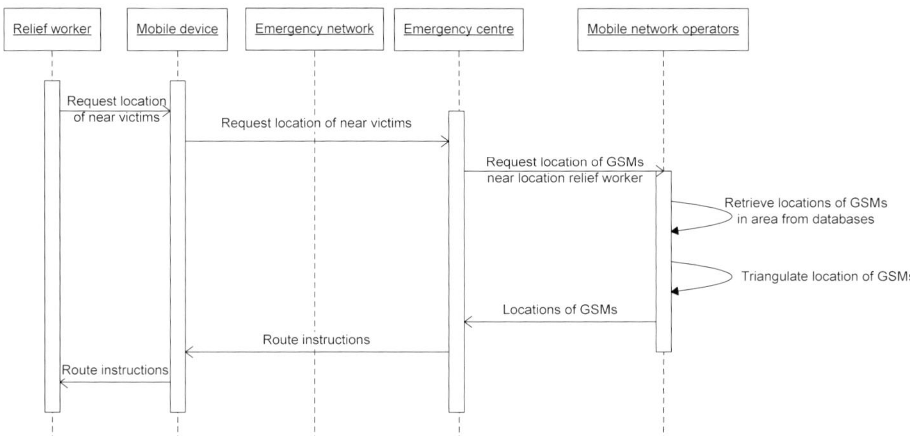

Figure 3-6 Sequence diagram of the 'Mobile satellite services' design example

The diagram shows that many actions have to be performed by the mobile network operators. This means that the success of the service system will depend on the level of collaboration between the operators, both from a technical and an organisational perspective.

3.9 Conclusions

Information and communication technologies have evolved enormously over the past decades. With the availability of high-speed communication networks, high-bandwidth mobile technology becomes available on the market for a low price. As a result, we are now able to develop service systems for anytime, anyplace usage. With the high speed in which technologies evolve, choosing the right technology is, however, not easy.

From a design perspective, a good technical design starts with a model of the interactions of the user and other actors with the service system. The most commonly used model nowadays is the use case diagram, as is further elaborated in Appendix A. The next step in the design process of a service system is retrieving the requirements for the technical architecture from these use cases. To get a better understanding of the technical architecture of a service system, we can then make use of UML sequence diagrams, as was shown in the case example in section 3.8. The example also stresses the importance of cooperation between organisations to make the service work. This organisational network will be the topic of the next chapter.

4 Organisational network

The organisational network deals with the interdependencies between resources in a (set of) service organisations. While many organisations continue to attempt to develop online or even mobile products and services on their own, increasingly the production and delivery of new services that meet market demands requires technological support that can no longer be provided by single organisations. Consequently, service concepts will be more focused on cross-organisational service provision through complex value networks. These attempts are hindered by inadequate organisational structures that impede the ability of agencies to tailor and deliver services. In this chapter we describe the creation of network organisations for mobile service systems, the business models, the actors/roles/member types, and changing relationships.

4.1 Blurring boundaries in mobile domain

The domain of mobile information services highlights the blurring of organisational and industry boundaries in the domains of telecommunications, IT and media. The roles needed to develop and deliver mobile information services are, for example, those of the mobile operator, service provider, content provider, hardware and software providers and, last but not least, the customer. Together they constitute a value network of actors who need to collaborate in order to deliver mobile information services.

These value networks differ depending on the sort of activities, which range from research on new radio technologies to contracting sales channels and everything in between. One factor creating these differences is the large differences in the time required to develop new network technologies and services. In the market it can be observed that while the innovation cycle for radio network technologies is approximately 10 years, for applications it is several years and for services only roughly 1 year. The involved actors, such as the operators, application developers and content providers, have to invest in the technology necessary to deliver their contribution to customer value. On the one hand this means that the risk is spread between different players but on the other hand it means that the different players are dependent on one another; and this is further complicated by the different innovation cycle periods. Furthermore, the revenue models will often not be established at the time investments have to be made.

These differences can be observed through actions, for example, in the Research and Development phase companies will collaborate in consortia such as OMA (http://www.openmobilealliance.org/) and work together with research institutes in research programmes that might be partly funded through innovation subsidies from government. In the service innovation phase companies operate in a network organisation and coordinate their design activities such as building prototypes and testing. In the daily business routine during the business development phase the value network is structured and governance mechanisms like a revenue model control the value network.

We do not expect this innovation cycle to be a linear process; rather we expect it to be iterative with many feedback loops.

Another distinction is between the exploration phase in which new practices are applied and the exploitation phase in which existing practices are applied. It is an assumption that networks for exploration can mainly be found in forms of cooperation that are usually labelled as Research and Development. This cooperation takes place long before market introduction and members make agreements on intellectual property issues. Often, the 'hard' R&D phase concerns new technology development. Also during the next phase when companies are collaborating on developing services the networks can still have a more exploratory character. We label this phase as Service Innovation. During this Service Innovation phase the network evolves from a more exploratory character into a more exploitation kind of network. In the phase in which the market introduction is prepared and the business case is built the network will be used for exploitation. We label this as the Business Development phase. This is illustrated in Figure 4-1.

Figure 4-1 also illustrates the position of the service system design phase in the design cycle: we position this between Research and Development on one side and Business Development on the other side. Thus, mobile information services are developed in a network where companies cooperate. First in the exploration phase in which network organisations design services by experimenting with and learning from using disruptive technologies and business models. If the service is on the market the network organisations successively exploit the service, this is the exploitation phase.

Nooteboom and Gilsing (2004) distinguish networks for exploration and for exploitation: exploitation entails improvements with respect to established practice, while exploration entails the development of new practices.

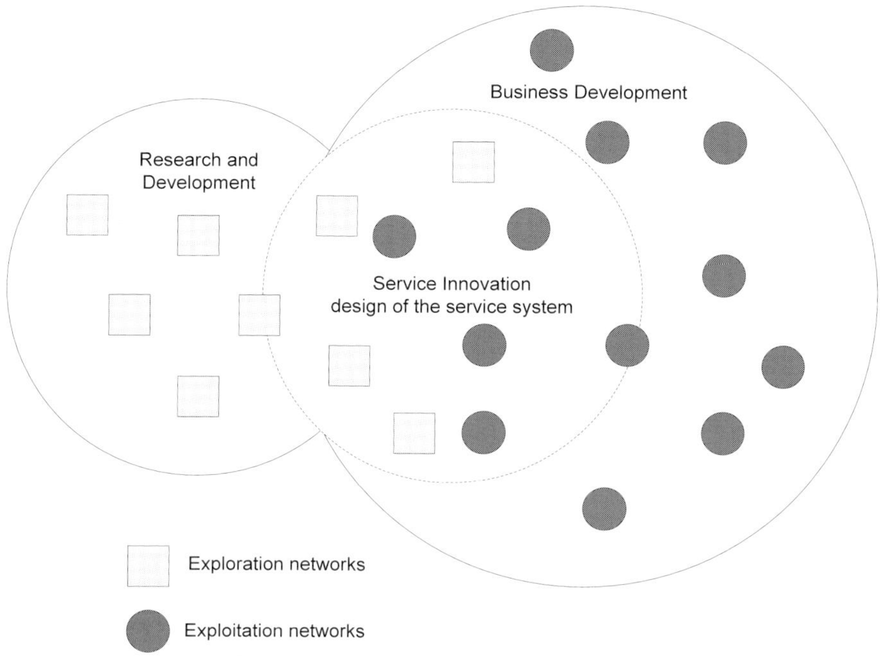

Figure 4-1 Positioning Service Innovation

Table 4-1 Networks for exploration and exploitation (Nooteboom & Gilsing, 2004)

Network features	**Exploration**	**Exploitation**
Network structure:		
Density	High	Low
Stability	Limited	High
Centrality	Low	Often high
Strength of ties:		
Scope	Wide	Narrow
Duration	Limited	Often long
Frequency of interaction	High	Low
Control	Low	High
Trust/openness	High	Generally low

During the design process of the service systems the network organisations might evolve from exploration towards exploitation orientation. This might happen in an evolutionary way or by starting again. In both situations, contracts and agreements need to be reconfirmed or renegotiated when a service will be exploited on the market.

4.2 Networks of organisations

In the logistic domain the concept of supply chains is well known. This concept was translated as value chains for a broader use (Porter, 1985). Value chains are helpful in identifying value-adding activities, but the complexity and radical changes in the merging telecommunication and media sectors are disregarded in these chains. In particular, the traditional telecommunications value chain is increasingly being deconstructed, which is leading to the development of a complex and rapidly evolving value network that consists of a myriad of network relationships (Li & Whalley, 2002). So currently, the rapid change in mobile telecommunication technologies, which in turn creates opportunities for new services, is increasing the interdependency of firms in the mobile sector. This increasing interdependency is one of the most frequently cited reasons for firms coming together to form cooperative relations (Gulati & Gargiulo, 1999). In addition to the resource interdependencies human resource constraints, compounded by time-to-market pressures, are driving the creation of inter-organisational networks for the provision of information services to mobile devices. While these networks are being formed to make innovative mobile services possible, they must also meet the requirements of a successful business venture. This can be quite a challenge. Thus, at the same time as mobile service networks are formed to overcome resource constraints they must also ensure that the overall venture is successful for the individual members while delivering highly valued services at the right price to the customers.

Powell (1990) begins his discussion of the network form of organisations by discussing why the familiar market-hierarchy continuum does not do justice to the notion of network forms of organisation. A starting point in this discussion is the transaction cost economics (TCE). TCE is an interdisciplinary approach to the study of economic organisation in which the transaction is made the basic unit of analysis (Williamson, 1985). Transaction costs serve as a measure of efficiency for the evaluation and selection of various institutional arrangements. Powell (1990) argues that besides economic exchange other forms of exchange also have to be taken into account. He compares the market, hierarchy and network form and describes them as follows.

- In market transactions the benefits to be exchanged are clearly specified, no trust is required, and agreements are bolstered by the power of legal sanction. The means of communication are prices. The degree of flexibility is high, the amount of commitment is low, the climate is precision and/or suspicion and the actors are independent.

- In hierarchies, communication occurs in the context of the employment contract. Relationships matter and previous interactions shape current ones, but the patterns and context of intra-organisational exchange are most strongly shaped by one's position within the formal hierarchical structure of authority. The means of communication are routines. The degree of flexibility is low, the amount of commitment is medium to high, the climate is formal, bureaucratic, and the actors are dependent.
- Network forms of exchange entail indefinite, sequential transactions within the context of a general pattern of interaction. Sanctions are typically normative rather than legal. The means of communication are relational. The degree of flexibility is medium, the amount of commitment is medium to high, the climate is open-ended, mutual benefits and the actors are interdependent.

Networks or network organisations have emerged as an organisational form to overcome the problems with hierarchies and to create greater structural effectiveness and responsiveness with trade-partners (Powell, 1990). These characteristics have enabled mobile carriers to use networks in order to access scarce new know-how and capital needed to develop mobile information services.

One way to describe an inter-organisational network is to use the characteristics of its members. If we consider a limited set of organisations, firms, this creates a more narrow set of network relations: the inter-firm network. Sydow and Windeler (1998) define an *inter-firm network* as an institutional arrangement among distinct but related profit organisations which is characterised by (1) a special kind of (network) relationship, (2) a certain degree of reflexivity and (3) a logic of exchange that operates differently from that of markets and hierarchies. The network relationship is typically complex and reciprocal and reflexivity implies problem solutions often requiring inter-firm coordination. The logic of exchange is one that combines cooperative and competitive elements, dependency and autonomy and trust and control.

A network must also have limits that define its membership. One such limit is the goal of the network, with network membership being established by each individual firm's contribution to the attainment of this goal (Jones, Hesterly, & Borgatti, 1997). Using this approach, a service network will be limited to those firms actively involved in the provision of the service.

Governance mechanisms are needed to control the emerging value network and its interwoven relationships between the different value chains. The need for governance stems from the special problems of adapting, coordinating and safeguarding exchanges created by networks of ties. It has been proposed that network forms of governance can overcome these problems by using social mechanisms rather than authority, bureaucratic rules, standardisation, or legal recourse (Jones et al., 1997). Given the network nature of mobile communications services and the position of the network operator in the industry, operators will act as the triggering entity, gathering partners to fulfil the roles necessary for provision of the service. These roles are filled in accordance with the evolving business model that defines the revenue model and the benefits to various members. We come back to business models in section 4.4.

4.3 Inter-firm service networks in the mobile industry

Currently, the mobile industry is undergoing a significant change as the network infrastructure is upgraded to next generation technologies. Previously, mobile operators, such as Verizon Wireless in the U.S., Vodafone in Europe, and NTT DoCoMo in Japan, were the dominant service providers, offering voice and text messaging, and controlling the customer relationship. While text messaging (SMS) sometimes brought relationships with third party service providers, these relationships were typically opaque to the end consumer. However, when new technologies made possible mobile information and entertainment services such as news and games, there was much speculation as to whether or not mobile operators would allow third party service providers access to the mobile networks and hence the mobile customers. Some speculated that operators would restrict access to their networks and create content themselves and some operators did indeed dabble in content generation activities. Currently, however, most operators have realised that a very broad range of applications and services can be developed for mobile and wireless devices, far more than they could ever create themselves, and thus have focused most of their attention on developing various arrangements for managing access to their networks. Therefore, the general structure of the mobile industry is evolving to one with literally hundreds of content and application developers offering services through mobile operators to consumers. Apart from the content and application developers and operators, such an operation requires the services of various intermediaries, platform developers, billing services, testing services, etc. The firms are arranged around the operators in elaborate network structures that due to the interdependencies have been labelled by the industry as the 'mobile ecosystem' (Natsuno, 2003).[2]

In this mobile ecosystem there exist multiple tiers of suppliers, similar to other industries that produce complex products. However, from the perspective of a mobile network operator, the inter-firm network consists mainly of a large number of content providers, with intermediaries that may fill a variety of roles in between (see Figure 1a). This configuration is similar to the ego-centric network, which consists of a set of direct, dyadic ties and the relationships between these ties, with the firm at the centre of the network (Hite & Hesterly, 2001; Powell & Smith-Doerr, 1994). We refer to this as the 'primary' service network and the business models developed by mobile carriers are primarily concerned with firms in this network.

However, if one considers the network to be all the firms involved with the provision of *one particular service*, the network looks a bit different (see Figure 1b). We refer to this as the 'secondary' service network. We contend that to understand the production of value, which in our case is service-specific, both primary and secondary networks should be considered. Furthermore, to understand the full extent to which firms from different industries (media, banking, advertising) are involved in the mobile industry secondary networks should also be considered.

---

[2] The concept of the mobile industry as an 'ecosystem' is not accepted by all. Some view the dominant business strategy as a closed model characterized by the operator's vertical integration of the production and distribution process, making them at the same time the provider of mobile services and internet access as well as the marketer of user handheld devices (e.g. Ramos et al. 2002). In either case, the operators maintain a dominant position in offering the services to the end-users and controlling the relationship with the consumer. Early use of the term 'business ecosystem' includes Moore's (1996) book titled 'Death of Competition: Leadership and Strategy in the Age of Business Ecosystems'.

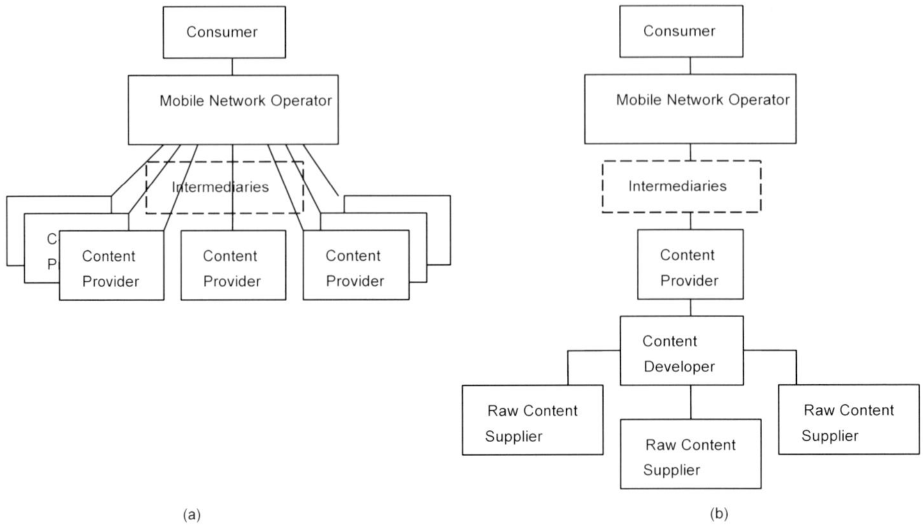

Figure 4-2 Inter-firm networks in the mobile industry

An example of a value network concept in the mobile domain is the iMode ecosystem. Natsuno (2003) describes the iMode ecosystem as a model centred on subscribers and content. In the middle is DoCoMo with a coordinating function that mediates between the subscribers and the content (Figure 4-2a). The coordinating role of DoCoMo in the ecosystem is to provide the business model, the marketing, servers (billing system), network and clients (mobile phones). The user experience is the critical factor in this complex ecosystem. iMode changed the traditional two-way business relationship between the service provider and the subscriber by adding a third player in the form of the content (or application) provider. iMode services are offered through an operator-run portal and are implemented through a business model that aims to establish a uniform set of governance mechanisms across a broad range of operator-content provider relations. An example of services delivered by a coordinating intermediary are the services of SMS brokers who deliver SMS voting services (e.g. vote for your favourite idol) via all available operators in a country to the public. The SMS broker intermediates between the content providers and the consumers (Figure 4-2b).

An important aspect of value is the value of network membership to each of its members. One motivation to join a value network might be generating revenue. However, in the 'MIES on the campus' service (see chapter 7) we found that most important motivations to participate in a value network are (a) working with partners with an established reputation in an innovative undertaking is that it supports a firm's image and (b) to learn and to achieve a competitive advantage. Partners will learn in the exploration phase from user tests and in the exploitation phase from market reactions.

4.4 Business models

While several definitions of business models exist (e.g., Afuah & Tucci, 2001; Amit & Zott, 2001; Hedman & Kalling, 2003; Weill & Vitale, 2001), one of the most frequently cited definitions of a business model is "the organisation (or 'architecture') of product, service and information flows and the sources of revenues and benefits for suppliers and customers" (Timmers, 1998). By this definition, the production of a particular product or service requires a network of firms and the components of a business model will define potential inducements for firms to create or join a network. Furthermore, the consideration of inducements for both suppliers and customers makes the business model concept attractive for analysing service networks. Thus, for IT-enabled services, examining the business model may provide insights into the inducements for network formation. In the case of the mobile telecommunications industry, technological advancements enabled the development of new services and they could not be produced and sold through existing networks. In order to produce and deliver new data services, including mobile games and mobile information services such as weather and stock reports, mobile network operators created new business models. While at first there was speculation that they would produce the content and applications required for the new services themselves, it soon became clear that service provision would require a network of service providers.

Therefore, examining the business models used to develop mobile services should provide insight into network formation. The business model components likely to play the greatest role in network formation are the revenue model and the benefits for suppliers and customers. The more detailed aspects such as the organisation of service and information flows may have operational and managerial implications but are less likely to influence formation. This proposition points to the dynamic nature of a business model. While research on 'idealised' business models implies a static and well-specified plan, in practice business models provide a general framework and continue to evolve as the firms that produce the service come together. Hence, while the general concept may exist a priori, it is adjusted as the network forms.

Returning to the definition, the business model consists in part of the revenue model, as well as supplier and customer benefits. The revenue model specifies the direction of revenue flows as well as how the revenues will be shared amongst the parties. It is surprising that something as basic as the direction in which revenue must flow would need to be specified but in areas of technological innovation it often unclear what value a particular component will have. For example, in the early days of mobile content and applications it was unclear whether network owners should pay content providers or vice versa. While overall revenue models tend to be highly idiosyncratic to the manner in which the service is produced and is reflected in the amount of specialisation and outsourcing, there are some standard models for the customer segment. These standard models include the subscription model, the advertisement model, the infomediary model, etc. (see Rappa www.digitalenterprise.org/models; see also chapter 3).

Factors influencing the revenue model include the need to recoup investments, the expected size of the customer base, as well as the number of firms required to provision the service. While a vertically integrated firm may have a good understanding of the possibility of recouping investments required for a new product, in network forms of production this process can be difficult.

In addition to the revenue model, a business model will also reflect the benefits to its members or suppliers. Research on inter-organisational networks has identified several benefits to member firms including acquiring resources, cost and risk sharing, market entry and learning, (Das & Teng, 2000; Porter, 1991). While this list is not exhaustive it does point to benefits that can be achieved through the framework presented in a business model.

The first of these, *acquiring resources*, has been labelled as the most oft-cited reason for firms to form alliances. This need is explained through the resource-based view (RBV) literature and explains a primary motivation for firms to seek to join a network. Business models, by laying out a plan for how the product or service will be produced, identify required actors and their roles. This is further explored in section 4.5.

The second supplier benefit is that of *sharing risks and costs*. By combining transaction cost economics with social network theories, Jones et al. (1997) proposed that firms would choose a network form of governance in response to exchange conditions that include demand uncertainty. The market for mobile applications has been very uncertain and under these conditions it is likely that firms would want to share risks, and hence costs of investing in technologies and developing business processes. These costs and the extent to which they are shared are laid out to some degree in the business model.

A third benefit that network membership may present is the *possibility of market entry*. Given the nature of the telecommunications industry, in which network operators control access to customers through handsets as well as through billing relationships, market entry for content providers is greatly facilitated by having a relationship with a network operator. Conversely, network operators who do not wish to develop their own content need to have relationships with a wide variety of content providers in order to attract customers and successfully enter the mobile content market.

The fourth supplier benefit to joining a network is *learning*. The research on networks and learning covers both how networks of firms learn (Powell, 1998; Powell, Koput, & Smith-Doerr, 1996) as well as how individual firms learn through participation in a network (J. E. Brown & Hendry, 1998). Here we are focused on the firm's expectation of learning as an inducement to enter the network. While a firm's ability to learn in the network will likely be dependent to a large extent on firm-specific factors, characteristics of the network as laid out in the business model may facilitate learning.

In addition the revenue model and supplier benefits, the business model must also specify the customer benefit. From the suppliers' perspective 'customer benefit' is often labelled as 'customer value.' From this perspective, network formation occurs when firms come together to develop and provide services based on their unique contribution to the end customer value. This concept assumes that there is some minimum level of agreement on what that customer value is; otherwise products could contain features that conflict. This agreed-upon notion of customer value may be determined by firms who are in a position to determine the dominant technical paradigm or through a more organic mutually agreed-upon perception of user value (Christensen & Rosenbloom, 1995).[3]

---

[3] The meaning of customer value in this context differs from those used by marketing and market researchers. Here service providers must *assume* what it is that customers value and hence could be referred as 'perceived customer value'. However, marketing researchers have done extensive studies on 'perceived customer value,' which refers to what consumers believe will be the value of a product *before* they buy it and is an important precursor to purchase. Therefore, to avoid confusion the term 'perceived customer value' while appropriate is not used. See chapter 3 for more information on perceived customer value.

Thus, with the revenue model, supplier benefits and customer value, the business model should provide a framework for understanding inducements for service network formation. However, theoretically the business model does not include a well known factor in alliance formation, namely embedded relational ties (Granovetter, 1985; Uzzi, 1996). While the business model, through its components of revenue model, supplier benefits and customer value, may be able to shed light on the types of firms that may desire to join a network, reducing the set of potential partners, the embedded relational ties variable often answers the question 'why this particular firm?' Therefore, in addition to the inducements presented by the business model, we add the opportunity of embedded relational ties as well as consider the type of formation process.

4.5 Actors, roles and resources

So, acquiring resources is one of the drivers for inter-organisational networks. People and firms need outside sources of cognition and competence to complement their own. Companies cooperate because they lack certain resources in their own company. (Nooteboom, 1999) argues that this is the fundamental reason why inter-firm linkages are important, especially for innovation. The role divisions vary, the revenue models differ and the benefits or value for being part of the network differ for each involved actor. Table 4-2 shows a standard list of roles for actors involved in mobile information services based on case study research in the domain of mobile information and entertainment services (Maitland, Van de Kar, Wehn de Montalvo, & Bouwman, 2005)

Table 4-2 Actors and roles (not exhaustive)

Role category	**Roles**	**Explanation**
Functionality related roles	Service provider	Provides billable service to the end consumer
	Network operator	Operates the mobile telecommunications network over which the data (service) is transmitted
	Platform provider	Provides the software that defines the general platform on which a variety of services are run
	Application provider	Provides the software that makes a service possible and that sits on top of the platform

Role category	Roles	Explanation
	Hosting/presence provider	Operates and maintains the server that hosts a website that is an integral part of the service, particularly relevant to the further development of content
Content related roles	Raw content supplier	Supplies content in a format unusable for the mobile service & terminal
	Content developer	Transforms raw content into content appropriate for the service and the mobile terminal
	Content provider	Provides 'appropriate'/transformed content to the service provider
	Content aggregator	Serves as an intermediary between the service provider and the content providers
Hardware roles	Equipment provider	Provides the hardware, physical components of network
	Handset supplier	Supplies platform or service-specific handsets
LBS roles	Content geocoding provider	Adds x/y coordinates to the content
	User positioning	Provides the position information of the mobile device
	Positioning technology vendor	Supplies user positioning equipment
	GIS provider	Provides geographical information, and GIS services, necessary to indicate location information of relevant content
Customer relation roles	Billing provider	Provides billing services to collect revenues from end consumers
	Marketing provider	Markets the service
	Customer support provider	Point of contact for customer queries regarding the service; responds to customer queries
	Content quality manager	Monitors and improves content quality
	Advertiser	Provides ads to user and provides income to the service provider
Customer	Customer	Buys the service
	User	Uses the service

Which roles have to be fulfilled depends on the kind of mobile information service. For example, the LBS roles are only relevant for location-based mobile services. This counts for both the exploration and exploitation phase.

In next subsection we discuss what the contribution from each role is and what this implies for the kind of partnership.

4.5.1 Resources and value networks

As the driving force behind network formation, resources are a crucial component of any value network and as discussed below will vary in terms of their fundamental characteristics, create interdependencies among network members, and give rise to power structures that can be used to delineate different types of partners inherent in a service network (Wehn de Montalvo, Van de Kar, & Maitland, 2005). Resources have been studied from many perspectives and the concept can be conceived very broadly to include almost everything in an organisational (capital, labour, infrastructure, technology, knowledge, processes, routines, capabilities) and inter-organisational setting (relationships, etc.). Distinctions can be made between *property-based* resources and *knowledge-based* resources, identified by Miller and Shamsie (1996). This can be compared with the distinction between *tangible* and *intangible* resources (Itami & Roehl, 1987; Wernerfelt, 1984).

Haanes and Fjeldstad (2000) identify tangible resources as concrete and tradable, such as factories, technology, capital, raw material, and land and intangible resources as difficult to transfer, such as skills, knowledge, relationships, culture, reputation, competencies. Essentially, this distinction parallels the two types of *property-based* and *knowledge-based* resources.

Building on this distinction, Das and Teng (2000) have identified three salient *characteristics* of resources and arrive at a matrix of resources that illustrates specific kinds of resources in each category. The basis for this classification is the reasoning that alliances need to be formed in order to obtain resources featuring imperfect mobility, imitability and substitutability. Imperfect mobility refers to the difficulty and cost of moving certain resources from one firm to another and obtaining them from the owner. Imperfect imitability and imperfect substitutability imply the difficulty of obtaining similar resources elsewhere. Complementing this with the external assets identified by Porter (1991) (reputation and relationships), (Wehn de Montalvo et al., 2005) arrive at the following illustration of resources in Table 4-3.

Table 4-3: Typical Resources; based on Das and Teng (2000, p. 42) and Porter (1991)

Resource Characteristics	**Resource Type**	
	Property-Based Resources (Tangible)	**Knowledge-Based Resources (Intangible)**
Imperfect Mobility	Human resources	Organisational resources (e.g. culture, reputation, relationships)
Imperfect Imitability	Patents, contracts, copyrights, trademarks, and registered designs	Technological and managerial resources, skills
Imperfect Substitutability	Physical resources	Technological and managerial resources, skills

The application of the resource-based view to research on strategic alliances provides the link with value network research in focus here. Strategic alliances can be regarded as a category of inter-organisational relations and networks. The common premise is that it is precisely the complementarity of resources that necessitates the formation and evolution of both, strategic alliances and value networks, and that none of the actors can make all the necessary components available for product development or service provision. *'The resource-based logic suggests that the competitive advantage of alliances is based on the effective integration of partner firms' valuable resources.'* (Das & Teng, 2000, p. 48). A resource-based perspective of the actors therefore provides a relevant basis to examine interdependence in the value network. From a resource-based perspective, paraphrasing Das and Teng (1998) on strategic alliances, *value networks are about combining resources that an individual firm cannot provide all on its own, yet are critical for the provision of a mobile service.*

4.5.2 Interdependencies

It has long been argued that all firms are embedded in one or more networks in which they collaborate with others to create value and in order to service the markets (Granovetter, 1985). Network boundaries are not easily defined because mostly there is no overarching purpose for the interactions. As shown in section 4.3, this is different for so-called value networks where the boundaries of the network can be more clearly distinguished by identifying the actors involved in the provision of a specific service. As opposed to a social network, in a value network the interaction among actors is goal-directed (i.e. the provision of a service) and cannot be assumed to be influenced merely by the individual actors' intention to influence each other. Value networks imply interdependencies (which may differ in their form and extent) among the organisations involved in it. Gadde et al (2003) have argued that each actor has a unique position in the network that is perceived differently by the different actors in the network because all have different relationships. The question is whether a more 'objective assessment' of the different actors' positions within the value network on the basis of the resources and their relevance or importance to value creation in a given network can be made.

> *'In social systems and social interactions, interdependence exists when one actor, does not entirely control all of the conditions necessary for the achievement of an action or for obtaining the outcome desired from the action. … Interdependence characterises the relationship between the agents creating an outcome, not the outcome itself.' (Pfeffer & Salancik, 1978, p. 40)*

Activities within the value network bring together different types of actors and resources and create (different) relationships of (inter)dependency. Theories on strategic management and resource dependence have often regarded interdependencies among organisations as inherently negative. Emphasis is therefore placed on how to manage interdependencies, on the implication of different coordination mechanisms (e.g. Ebers, 1999) and on strategies to restructure the conditions of interdependence (e.g. Mintzberg, 1979, 1983; Nassimbeni, 1998). The interdependence between various relationships in the network implies that a certain actor's change in behaviour also influences the position of other actors (Axelsson, 1987).

Several forces are identified binding networks together (Håkansson, 1987):

1. functional interdependence (actors, activities and resources as a system that is functionally related),
2. power structure (actor power based on activities and resources),
3. knowledge structure (activities' design and resource use bound together by actors' knowledge), and
4. time-related structure (network as a product of its history).

For an analysis of the strategic position of actors within the value network in terms of their contribution to value creation, the second type (power structure) is of greatest interest. Actor power is assumed to be based on the activities and resources of a particular actor. In particular, we argue, that actor power stems from the characteristics (degree of mobility, imitability and substitutability) of the resources. To typify the power 'structure' among the actors in a value network, (Wehn de Montalvo et al., 2005) propose a distinction between *essential*, *network-specific* and *generic* resource contributions to value creation, ranging from greater to lesser relevance to value creation in the network and based on resource characteristics, defined as follows.

- *Essential* resources as resources that are indispensable to the value network and the service it provides. These resources cannot be replaced without affecting the existence of the service and they are highly immobile, difficult to imitate or substitute.
- *Network-specific* resources are crucial for the service that the value network provides, yet their replacement would be possible without affecting the service directly. They are fairly mobile, possible to be imitated or substituted.
- *Generic* resources are required for the provision of the service but they are so general that they could be replaced fairly easily without impacting the service. They are reasonably mobile, imitable or substitutable.

This distinction provides a basis to define different partner types in the value network: *structural*, *contributing* and *supporting* partners (ranging from greater to lesser actor power depending on the kind of resources they contribute), thus identifying the nature of interdependencies in a given network and the strategic position of actors within the network (Ballon & Hawkins, 2003). These network partners can also be described as members of a network.

4.5.3 Summary of members types

Members in a network are actors that provide roles, see examples given in Table 4-2. The roles they provide consist of property-based and knowledge-based resources. The distinction of different resource contributions provides a basis to label different member types in each value network: *structural members* provide essential resources; *contributing members* add network-specific resources; and *supporting members* contribute generic resources to the process of value creation. This allows identifying the nature of resource-based interdependencies in a given network and the strategic position of actors within the network.

To capture the importance or relevance of different resources to value creation in a given network, a distinction between *essential*, *network-specific* and *generic* resource contributions to the value network can be made and each of these contributions may be in the form of tangible (property-based) or intangible (knowledge-based) resources. For analysis, the following matrix can be used to 'map out' the different resources in a given value network and their relevance or importance to it.

Table 4-4: Member types and resource contributions in a value network (Wehn de Montalvo et al., 2005)

Member type	Resource contribution to value network		Actor
structural	essential	tangible	?
		intangible	?
contributing	network-specific	tangible	?
		intangible	?
supporting	generic	tangible	?
		intangible	?

At this level of analysis, it will be possible to carry out a cross-case comparison of power structures in different value networks and identify similarities and differences in the positions of the various actors. This can be done in retrospective when analysing network organisations that explore or exploit services. In the following examples designs of services are described including the needed roles to provide those services. The structural, contributing and supporting member types might be foreseen however they can only really be distinguished when a network organisation is in place.

4.6 Design example: Last minute business travel changes service

This is a design example for the travel domain. The service in this case provides a set of interrelated services for business travellers who have to make last-minute changes to their travelling schedule. The focus is on a business traveller who is on a business trip while something unplanned takes place, which generates conflicts in his schedule.

First, the service concept of the service system is briefly introduced. Next, the organisational network of the service system is described in depth.

4.6.1 Service concept

Frequent business travellers from large corporations often find themselves in situations where they have to make changes to their previously planned schedules. Even a small schedule change can affect previous flight, hotel and taxi bookings. Changes to the travelling schedule involve a lot of work. Bookings have to be cancelled, and new alternatives have to be generated and chosen. This is a complicated and iterative process because current booking services in the travelling industry are independent of each other, making last minute changes a boring and time-consuming activity.

Business travellers often lack the time to make the changes themselves, as they will often have long and intensive business meetings. The natural reaction of business travellers under such circumstances is to contact their own organisation or a travel agency. This has also disadvantages, for example when the business traveller is in another time zone and the resources are not available at their own organisation to make the changes. The business traveller will also keep worrying that the process of changing his schedule will not be successful.

The service in this design example will minimise the inconveniences that are currently encountered by business travellers. The main idea behind the service is that it takes care of the entire process of changing the schedule. To do so, the service provider bundles the booking services that currently exist in the travel industry. The service will use these services to make a new schedule for the business traveller. In this process, it takes the appointments of the traveller and his preferences into account.

The service serves as insurance for the business traveller that the changes will be made correctly. In that way, it will increase the productivity of business travellers. They can now concentrate more on their business objectives, instead of being worried about their planning. We also expect that this will give business travellers more satisfaction in their work.

4.6.2 Organisational Design

Because the service combines several existing travel services, many actors will be needed to make the service work. Figure 4-1 provides an overview of the value network of the service system.

The roles, activities, responsibilities and benefits of the relevant actors are described below.

Business traveller

The business traveller is the customer who uses the service. The business traveller will benefit from the service as it makes the process of last minute travelling changes easier and less time consuming. The business traveller will only need to have interaction with the provider of the service.

Business organisation

This is the organisation that employs the business traveller. The business organisation will have interaction with the service provider, and also with the billing provider who will send the bill for the changes that are made. The business organisation will benefit from the service by increased productivity and efficiency of the business traveller.

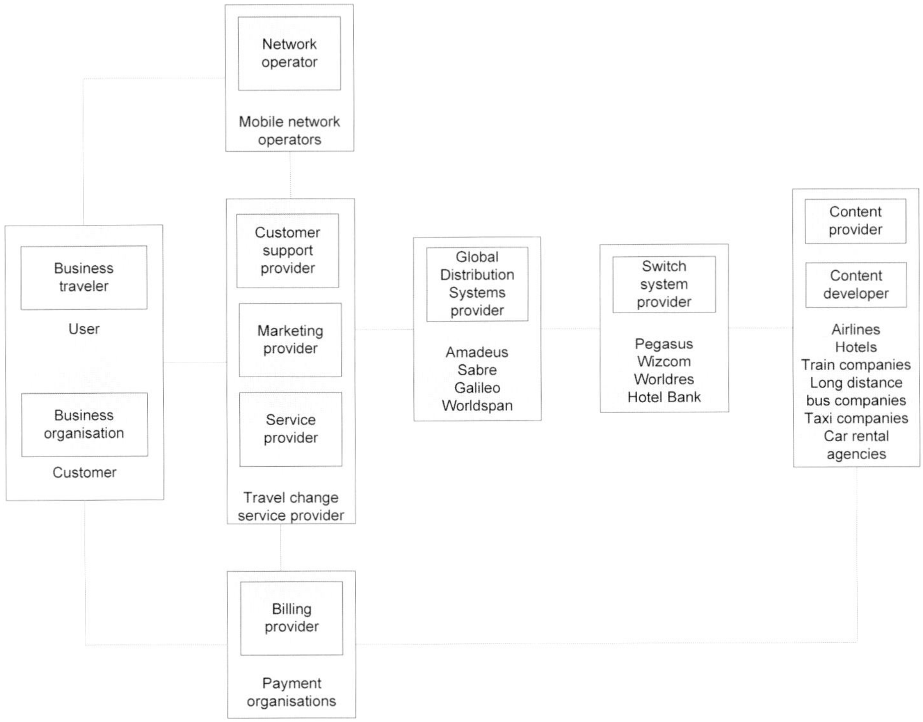

Figure 4-3: Value Network for last minute business travel changes service

Travel change service provider

This new organisation will provide the service to the business travellers. The organisation is responsible for the entire operation and reliability of transactions made regarding changes in reservations. It is also responsible for managing the dynamic information that is part of the profiles of customers such as the personal preferences and reservations of the customers. The service provider will also perform marketing for the service, and answer questions from customers.

The service provider will benefit from the service by charging the business organisations for using the service.

Mobile network operators

This actor will enable communication between the business traveller and the service provider. Business travellers will typically be on the move, and therefore wireless access to the service will be needed. The network operators will benefit from the service as they can charge the business traveller for using their networks.

Global Distribution Systems (GDSs) providers

A GDS is defined as following: "a computer distribution system for displaying available services, affecting bookings, and ticketing by tourism producers – airlines or otherwise – on an international scale". There are four major GDS providers: Amadeus, Galileo, Sabre and Worldspan. These systems essentially provide travel agents with the ability to make airline ticket reservations directly into the airline computer reservation systems. GDSs will thus play an integrative role in our service by contributing to the automation of activities involving the rescheduling of flight reservations. The activities of the systems involve collecting, integrating and distributing the content from content providers to other parties like travel agents. Maintenance and updates of the system are also part of their activities.

Switch system providers

Although GDSs in the past only handled airline reservations, it soon became apparent that they would also have to support other travel services such as hotel chains and car rental agencies. Therefore, 'switch-systems' were designed to connect reservation systems to hotel chains, thereby making them available through GDSs. The role of these switch-systems is therefore to enable the flow of content from various travel services into a GDS. It is a significant component for our service since it is necessary to link various travel services with one another to enable triggers between different reservations systems.

The GDSs and switch-systems can benefit from the fact that through the introduction of the service, they can achieve wider application of their systems. Furthermore, these systems that participate in delivering the service can achieve a better competitive position since it enhances their own system and the service they offered. GDSs and switch-systems generate revenues through charging the other actors in the value network.

Content providers

The content providers are the airlines, hotels, taxi companies, car rental agencies, train operators, long-distance bus companies etc. Their role is to constantly provide accurate and up to date information regarding their travel services. These organisations are also responsible for providing information regarding the profiles of customers, for instance the in-flight or hotel room preferences or the recent flight/hotel history of a traveller to the service provider.

The content providers are able to benefit from the fact that they can differentiate themselves from competitors who do not participate in this project. When the customer is provided with a list of flights or hotel rooms, the participating airlines and hotel chains will be considered and perhaps chosen by the customer.

Billing provider

The role of the billing provider is to coordinate the billing and payment of different organisations for delivering the service. It is responsible for ensuring reliable and secure payment between different parties. In order to fulfil their role, billing providers will obtain and utilise payment services that are available on the market.

The billing providers will benefit through the increased number of transactions that take place between business organisations and other parties in the value network. Billing providers will probably charge per transaction that takes place.

4.6.3 Result

This design example shows that many actors are involved in a service that looks relatively easy: making a change in a business trip schedule. This provides opportunities for new intermediaries who can use ICT to work efficiently. By choosing for a network organisation form these new intermediaries can start their business in a complex organisational landscape.

4.7 Design example: People Tracing Service

This is a design example for the emergency domain. It focuses on victims of disasters at airports. Disasters can happen on an airport, such as the collapse of a terminal or a plane crash. When this sort of thing happens, the family and friends of people who are at the airport try to find out whether their loved ones were involved in the disaster. Currently, there is no effective and efficient system to provide unambiguous and fast information about the location of victims. The objective of this design example is to design a service that helps people to find the location of a close-one who might have been involved in a disaster at Schiphol airport. The design of the service is based on Schiphol Amsterdam, but could also be a starting point for designs for other airports. In this case, we will first briefly introduce the service concept of this service system, and then emphasise the organisational network of the service system.

4.7.1 Service concept

An airport is a place where many lives may be in jeopardy. It contains a lot of people in a small area and aeroplanes take off and, loaded with thousands of gallons of kerosene. Fortunately there are a lot of safety measures and protocols that should minimise the possibility of a disaster happening. However, disasters still occur from time to time.

During a disaster, many different teams are involved in rescue operations and are trying to get the situation under control. In this hectic scene people like family and friends at home may be trying to find out if some one they know of was of involved in the accident and if so, where they can find them. People who are at the scene of the disaster and who are not heavily injured might want to search and find out where lost friends or travelling-companions are. Because of the high working pressure on the medical teams, there is no effective and efficient way to provide clear information about the location of these victims quickly.

The People Tracking Service (PTS) will help to find persons that might be involved in the disaster at Schiphol airport. It will combine information from hospitals near the airport, airline companies, check-in desks and information from ground personnel. All information about the whereabouts of people and their personal information (name, gender, etc.) will be combined and presented to next of line. The service can be used by people who are at Schiphol airport, and by people at home.

The service will provide information about which people have been taken to the hospital or to a save of area on TV screens or on digital billboards at the airport. The information will also be available from ground personnel who will form a so-called `PTS-team' and who are available to give information. At the Information Access Centre (IAC), which is an information point for passengers, mobile computer terminals will be available for accessing the service.

People at home who are worried that a relative is involved in the disaster at the airport can access the service on a website. On the website, a user can fill in the name of a person and then search for him. The service will then present information about the state of the person and his location.

4.7.2 The organisational network

Who are the actors and which roles do they fulfil? What is the benefit for each of the involved actors?

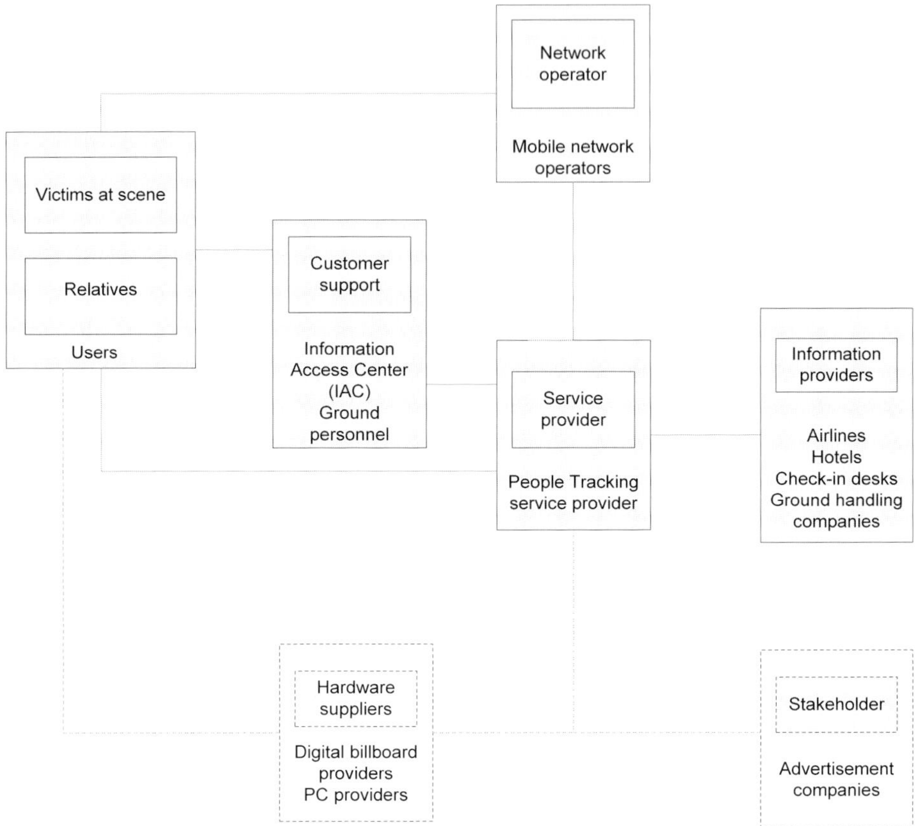

Figure 4-4 Value network for people tracing service

Hospitals

The hospitals near Schiphol airport play a vital role in achieving a high performance of the PTS. The PTS service system should constantly be updated with data from the hospital's databases. Especially for those people that are in a critical condition, it is important to fill in data about them in the system with high priority.

The big advantage for hospitals when the PTS works correctly is the major reduction in incoming phone-calls from worried people trying to find someone. They will now look at the website for information.

Airline companies

The airline companies that arrive and depart from Schiphol airport have to enter the information from their passenger lists for each flight. In that way, in case of an accident they can be traced faster with the PTS system.

Information Access Centre (IAC)

The IAC will connect to the PTS website in the event of a disaster. People could then come to the IAC and access the service system. The IAC could also enter new information in the PTS system, for example when a missing person turns up at the information point.

Adverting companies at Schiphol airport

In the event of a disaster, the adverting companies have to remove their advertisements from the digital billboards, and instead show the information from the PTS system. By cooperating in this, the adverting companies show their goodwill for the right cause, which creates positive exposure.

Mobile network operators

The PTS-team will access the PTS system to provide information to the people at the airport. To do so, they need wireless connections to the system. Preferably, the mobile network operators should reserve some capacity in their network for the PTS service in case of a disaster.

Schiphol Airport ground handling companies

At Schiphol, a small group of companies is active in checking-in passengers at the front desk. When travellers check-in at the front desk, their information will be entered in a database. This information has to be sent to the PTS system. Active participation of the ground personnel is not needed here. The PTS system will automatically send requests and retrieve information from the database itself.

The ground personnel will have an active role in the event of a disaster happening, because they will be (maybe not all) used as member of the PTS-team.

Users

The active participation of users should be minimised because they will obviously be in a state of panic, stressed, and uncertainty. Therefore the PTS website should be very clear and simple. The users only have to click, enter a name and read the results from the screen.

4.7.3 Result

This design example describes the Passenger Tracking Service, which enables victims to find their friends or relatives at the scene of a major disaster at Schiphol airport. The service uses information from various sources, including airlines, airports, hospitals and the victims and relatives themselves. The design example presents the organisational network that is needed for the service, including the actors involved in providing the information, their roles and their relationships. In Figure 4-4 it is visualised that some actors are more structural than others by using dotted lines for the supporting actors.

4.8 Conclusions

The design of mobile service systems increasingly demands resources that cannot be provided by one organisation. This means that the organisation has to cooperate with other organisations. The network organisation form seems more suitable for mobile services than hierarchy or market transaction oriented organisation forms. These network organisations are often quite complex and the members of these network organisations will fulfil different roles which are more or less structural, contributing or supporting in the network. The design process in which these networks are formed also has to be designed itself. In next chapter we also describe theories that support the 'soft' system design that includes actors and their (potentially) conflicting interests.

5 Design approaches for Service Systems

In the past three chapters, the characteristic features of services and service systems have been discussed. The focus in these chapters was on *what* mobile services systems are. In this section, we present approaches for designing mobile service systems, thereby shifting our focus towards *how* to design services systems. Various requirements on how to design these systems have emerged from discussing the specific characteristics of services systems. In this chapter several schools of design approaches will be presented and their suitability for designing mobile service systems is evaluated. We will show how certain of these existing approaches can be adopted for the design of mobile service systems. Besides the usage of design approach elements from longer existing disciplines, more and more results from research specifically focussed on service innovation and mobile service development is being published. We refer to these recent findings, such as the New Service Design (NSD) literature as mentioned in chapter 1, when appropriate.

This book deals with services and services systems that are enabled by information technology. So, naturally, we should be able to find inspiration from systems engineering which we do in section 5.1. We further deepen our knowledge of specific areas of system engineering: product design (5.2), information system design (5.3), and collaborative business engineering (5.4). Section 5.5 explores how modelling can help in service system design. We especially look into the possibilities of component based design and the use of building blocks.

The design, implementation and use of mobile service systems involves many stakeholders, such as providers, operators and users who together form a value network. These stakeholders may have different ideas, views, interests, and means with regard to the services that are designed. In section 5.6 we look into the design of services from a user's perspective. During the design process with the other actors cooperating in a network, decision making is an inseparable part of designing, which is complex and needs to be managed. Process management is complementary to the more traditional project management techniques that are also needed in the design of mobile services systems. In section 5.7 process management and project management are discussed as ways to deal with management and decision making issues in designing mobile service systems.

Finally, the principles, aspects, elements and techniques that can be useful for the design of mobile service systems are presented together in section 5.8.

5.1 System engineering

Considering the complexity of mobile service systems, with the many disciplines needed, the many interrelated systems components that need to be considered, and the dynamic environment in which mobile service systems need to be designed, a systems approach is needed. Systems' thinking (Checkland, 1993; Weinberg, 1975) includes holistic thinking, i.e., understanding the system as a whole in time (the system as it develops through its life cycle from conception until disposal or recycling) and space (a whole separated from the environment). All the components and organisations involved in a service-oriented system have to work together to provide a mobile service that adds value to the users. Mobile service systems are large scale technology enabled systems. For the design of such systems multiple disciplines need to be deployed and multiples stakeholders need to be involved. Systems' thinking also implies understanding dynamics and feedback relations between all these system components and between the system and its environment.

Engineering, as in systems engineering, can be defined as creating cost-effective solutions to practical problems by applying scientific knowledge to building things in the service of mankind (Maier & Rechtin, 2002; Shaw & Garlan, 1996). Essential in many definitions of engineering is the notion of solving real world problems using scientific knowledge. In our view, design is included in engineering, as one of the phases in the systems' life cycle.

Sage and Armstrong (2000) present a list of important issues that must be dealt with to engineer large systems. These issues are (1) many considerations and interrelations; (2) many different and perhaps controversial value judgments; (3) knowledge from several disciplines; (4) knowledge at the levels of principles, practices and perspectives; (5) considerations involving product definition, development and deployment; (6) considerations that cut across three different life cycles associated with systems planning and marketing, RDT&E, and system acquisition or production; (7) risks and uncertainties involving future events that are difficult to predict; (8) fragmented decision-making structures; (9) human and organisational needs perspectives and value perspectives as well as technology perspectives; and (10) resolution of issues at the level of institutions and values as well as the level of symptoms.

Each of these issues needs to be dealt with when designing mobile service systems. Some examples will illustrate this. The mobile field is characterised by uncertainties (issue 7); for mobile service systems these uncertainties include technology, the demand, the supply market and the interaction between demand and supply. This shows that mobile services have to embrace the perspective of the need of the customer, the customer value, the value for the involved business actors delivering the services and the technology perspective taking into account the dynamic development in the mobile field (issue 9). Decision-making takes place at a company level, at a dyad, triad and network level. This makes decision-making complex and potentially fragmented (issue 8).

It is clear from the above that in the design of mobile service systems we need to deal with these issues. We should therefore consider systems engineering approaches as a basis for a design method for (mobile) service systems, and look both at hard systems engineering and soft systems engineering. The hard systems approach departs from the idea that the present state and desired future state can be fully described, and are known. There is a strong use of scientific methods and quantitative modelling, and the aim is to find the optimal design or solution for the system that is being studied. Important schools are operations research (Winston, 2004), Systems Engineering and Analysis (Blanchard & Fabrycky, 2006; Buede, 2000; Sage & Armstrong, 2000), and System Dynamics (Forrester, 1968). Hard systems thinking can, however, not deal very well with the 'human element' or with qualitative factors, which are so important in service systems. Soft systems approaches, which were developed in the late 1980s and 1990s (Checkland, 1993; Checkland & Holwell, 1998; Checkland & Scholes, 1990) take different worldviews, qualitative factors, and the human element in the system as a starting point. Debate and reaching agreement among the different stakeholders are considered to be more important than the 'one outcome' of the hard systems approach.

As humans have been designing artefacts for a long time, many different design methods exist from which we can learn for designing service systems. Examples are methods for product design and methods for information systems design. Both product design and information systems design have their own literature fields. We briefly explore these design theories for products in section 5.2 and for information systems in section 5.3. Then, we will use a number of components from product design, information systems design, and hard and soft systems thinking, and describe an integrated method that is sometimes called 'business systems engineering' (Van Meel,

1994) or 'collaborative business engineering' (Den Hengst & De Vreede, 2004).

5.2 Product design methodologies

Product design is important for service design for two reasons: (1) there is a vast body of knowledge with respect to the design of commercially and technologically feasible products and (2) the clear distinction between services and physical goods seems to be rapidly eroding. Services are more often than not supported by tangible artefacts, such as the mobile phone that comes with the cell phone subscription. On the other hand, tangible artefacts (products) gain more added values through add-on services, such as a 24 hour helpdesk for a home computer.

Products are technical artefacts for which product design theories exist. Product design is about the prescription of structure, construction and use of a physical entity. Generally, product design is executed without the explicit order of a client who will also be the user of the product. Commissioners of the design project typically are the marketing departments of organisations. A typical feature of product design is therefore the uncertainty with regard to the user requirements. As a result in product design approaches great attention should be paid to revealing the needs and wants of the user, as the ultimate success of the product depends upon whether the end consumer will buy and use it (Rouse, 1991). User wants and needs are translated into explicit specifications suitable for the technical designers. These specifications are the touchstone for the whole project that follows.

Generating an idea for a product is part of an innovation process with divergent and convergent activities. Product planning is part of the strategy process of a company, and the realisation of the product consists of a technical development process and a commercial development process. (Roozenburg & Eekels, 1995, referring to Hall 1969) argue that the basic product design cycle within product development is quite close to a special implementation of the empirical cycle of the problem-solving model of systems engineering. They present several product design overviews. Their basic cycle of design is presented in Figure 5-1.

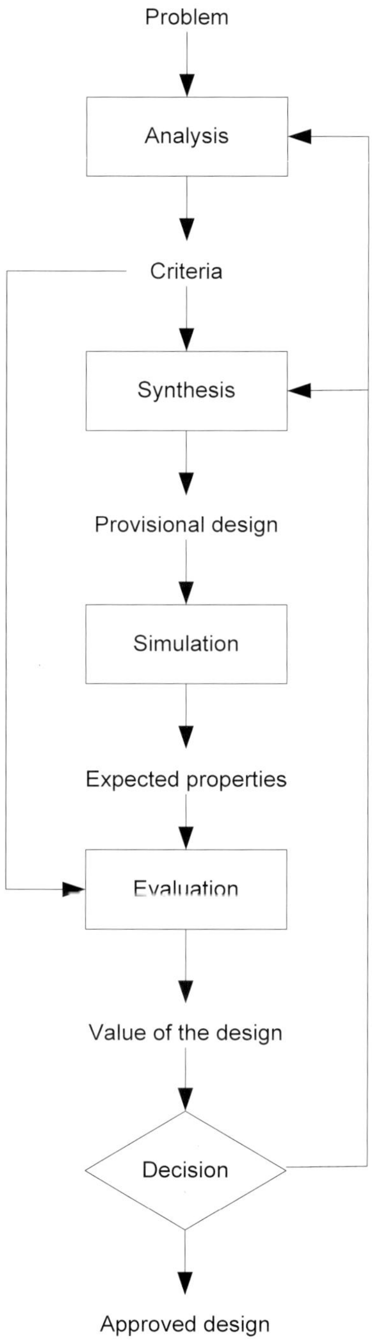

Figure 5-1 The basic cycle of design (Roozenburg & Eekels, 1995)

The basic design cycle provides valuable input for the design approach for mobile service systems. Additionally, (Roozenburg & Eekels, 1995), provide methods, rules and recommendations for each phase of the basic cycle. Although these are directed towards tangible products like vending machines and phones, some of these methods, rules and recommendations provide inspiration for designing services. (Buijs & Valkenburg, 1996) present a similar kind of model for developing innovative products with the phases: analysis, synthesis, materialisation and optimisation. For each phase they present the input, activities and output with an evaluation between each phase.

Roozenburg and Eekels refer to Quality Function Deployment (QFD) in their description of how to get from a design problem to a design specification. QFD is a method for structured product planning and development that enables a development team to specify clearly the customer's wants and needs, and then to evaluate each proposed product or service capability systematically in terms of its impact on meeting those needs (Cohen, 1995). The method is tailored to the organisational context in which the process of developing a specification takes place (Roozenburg & Eekels, 1995). In QFD the sequential approach of product development is replaced by a concentric, concurrent approach in which different stakeholders are involved in the product development process from beginning to end. The method supports a team of designers in identifying the needs of the customer and interpreting them in terms of technical parameters and target values. The QFD process involves structuring matrices of which the first displays the customers' wants and needs. This matrix is the most well-known and is called the 'house of quality'. In QFD handbooks the activities are structured in phases: phase 0 = planning phase; phase 1 = gather the voice of the customer; phase 2 = build the house of quality; phase 3 = analyse and interpret.

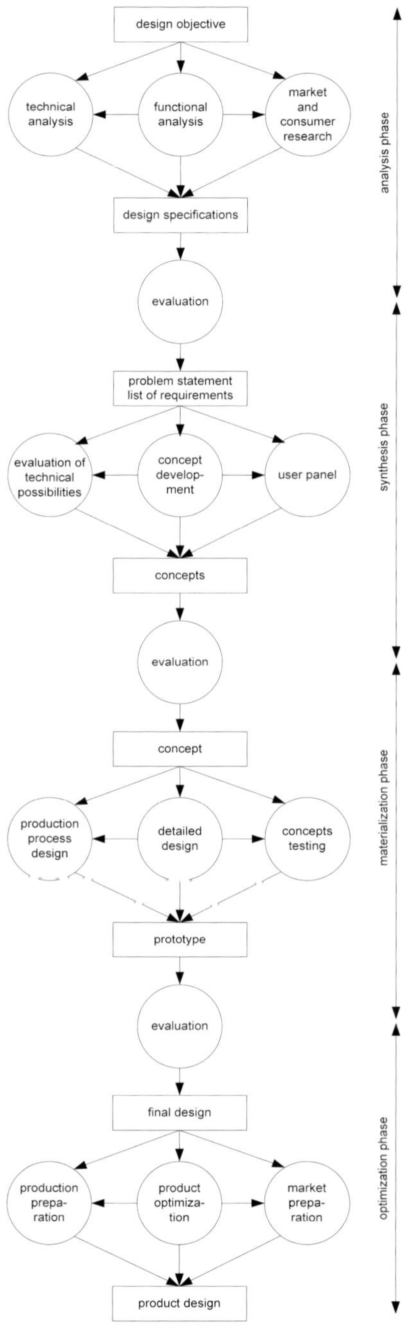

Figure 5-2: Model of the development phase of the innovation phases model (translated from Buijs & Valkenburg, 1996)

Creveling et al. (2003) present a methodology for product and technology development which is built upon a balanced portfolio of tools and best practices and enables a product development team to develop the right data. Their methodology is based on the Six Sigma Approach created by Motorola and popularised in corporate business by General Electric. In their view, the critical components in product development are (1) the product development process; (2) tools and best practices and (3) project management. Creveling et al. present a technology development roadmap (invent/innovate, develop, optimise, verify), which deals with the manufacturing system, and a product development roadmap (concept, design, optimise and verify). They link quality function deployment (QFD) to the invent/innovate phase in the technology roadmap and the concept and design phases to the product development roadmap.

Product design is about the prescription of structure, construction and use of a physical entity. Service designers can adapt this prescription for the design of intangible goods: services. Like product designers, service designers have to deal with the conflict between the need for creativity and uncertainty reduction. One might wonder why the established product design methodologies are not more accepted for service design. The question 'how does service design differ from product design' is often asked. The nature of services demands that its design tools support the characteristics of services as described in Table 2-1 in chapter 2; However, some design tools can be used for products as well as service development. Verkuijl, Tischner and Tukker (2006) present an overview of product-service development tools and show that many of these tools can be used for products as well as services. We agree and like to emphasize that scenario development methods in which the user and his/her context are taken as a starting point can be used for products as well as services. Especially for services since the 'the act of the value is gone with the consumption' and therefore the service development has to start with the viewpoint of the user; the user's experience of the delivered service is integral part of the service. This also counts for ICT enabled services which offer only value when of current interest; e.g. a location based service has only value if the service uses the current position of the user and not the position of a few hours earlier; the weather news is only valuable if it is up-to-date. The service to be designed is most likely part of a multi-channel strategy of a company. When defining a multi-channel service mix, focus and progression in the design process benefit from defining the e-services first and then adding multi-channel services (Simons, 2006). (Simons, 2006) developed the MuCh-QFD method to design e-services in a multi-channel context based on QFD.

Next section shows what specific information system design approaches are useful for services.

5.3 Design of information systems

Alter (2002) defined information systems as "a work system whose business process is devoted to capturing, transmitting, storing, retrieving, manipulating and displaying information, thereby supporting other work systems". It is very clear that the types of mobile service systems that we have described in earlier chapters and in the examples at the end of each chapter are examples of information systems. Information from the users is captured with a mobile device, transmitted to back-end systems, stored for later use, manipulated and combined with other information, and presented again on mobile devices after transmission. Mobile service systems belong to a category of information systems that are hard to design and test, because they are distributed (Tagg & Freyberg, 1997). The importance of communication and interaction with the users also need more emphasis than when designing 'classical' information systems.

Waterfall and spiral models

The waterfall and the spiral model are two basic development models for information systems (Cadle & Yeates, 2001). The waterfall model originates from the seventies, while Boehm (1988) introduced the spiral model at the end of the eighties. The waterfall model is a stage-by-stage model, the spiral model combines this with an evolutionary approach. The aim of the spiral model is to give a structured and incremental approach to the service design process. The project starts at the centre of the spiral and progresses outwards, so problems occur during the various stages and are not all found at the end. In three or more rounds various quadrants activities are carried out until the service designers are satisfied with the results, see Figure 5-3.

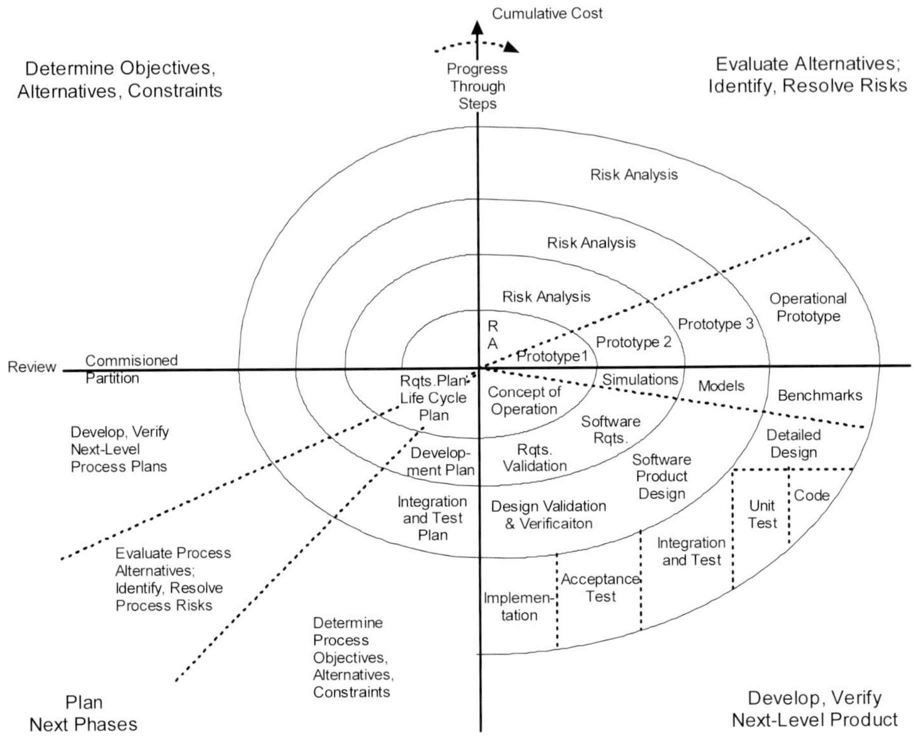

Figure 5-3 Original diagram of spiral development (Boehm, 1988)

Evolutionary design

It is difficult for users to list requirements for information systems in which they are involved. The same applies for mobile services. Different users will have different requirements and requirements will change over time. This problem can be mitigated through the inclusion of users in the design phases of mobile services as in an evolutionary development model. An evolutionary approach is suitable in situations where the requirements are not well formed or understood by the users, where it is difficult to specify the requirements, or where it is difficult to determine how a proposed solution will perform in practice. "An evolutionary design approach consists of expanding increments of an operational product with the directions of evolution being determined by operational experience.... the disadvantage of this approach is to distinguish it from the old code-and-fix model, whose spaghetti code and lack of planning were the initial motivation for the waterfall model" (Boehm, 1988).

Often information systems are part of service systems and therefore we can use some elements from the information systems engineering field. A spiral model for (information) system development seems to fit well with the design of mobile service systems, as it more or less provides some structure to the evolutionary design procedure that is needed for mobile service systems design. Design needs to balance incremental change and radical thinking. It is striking that right balance asks for continuous input from different stakeholders during the service system design process. Therefore, the next section introduces a collaborative design approach that combines elements from hard systems thinking and soft systems thinking.

5.4 Collaborative Business Engineering

Designing mobile service systems without considering the organisational structures and processes, i.e., the value network and the operational processes, might result in a design that is neither effective nor efficient, given the existing organisational processes. The Collaborative Business Engineering (CBE) approach focuses on collaboration within business processes. An organisation cannot design mobile service systems on its own, since these systems are the result of coordinated design activities by different actors like operators, handset manufacturers and content providers. Participation of various experts and stakeholders with much tacit knowledge of the situation is needed. During the engineering process stakeholders and experts need to work closely together, for instance in workshops and brainstorming sessions. The 'collaborative' part is also important for this research since mobile services have to be designed in a network of actors, as discussed in chapter 4.

Den Hengst and De Vreede (2004) give an account of a number of case studies in which a collaborative design approach has been applied. They describe these case experiences and formulate the lessons learned that should ultimately evolve towards what they call, Collaborative Business Engineering (CBE). CBE is a good source of inspiration for a design approach for mobile service systems for several reasons.

The CBE approach aims to deal with ill-structured problems for which bounded rationality is seen as the dominant decision-making paradigm. The design approach should not focus on finding the single optimal new organisational design, but on facilitating a diagnosis and design process that will yield a satisfying and acceptable solution (Den Hengst & De Vreede, 2004), which is analogous with the soft systems thinking approach described earlier in this chapter.

The CBE approach is a systems engineering approach in the sense that it aims to establish changes in complex systems. They are problem driven in the sense that the design, engineering or change process is initiated and motivated by problems or opportunities perceived by a group of stakeholders. For the design of mobile service systems opportunity driven is usually more appropriate since the design of mobile services is more often aimed at finding new business for companies than at solving problems for the users. The 'opportunity' owner wants to make a profit with new mobile services and has to take a customer driven or market driven approach in a market with a lot of uncertainties. The services will only be successful if they are useful and help to solve problems and meet the needs of the (potential) users.

A purely hard systems engineering approach would not suffice in view of the involvement of many stakeholders, the ambiguity of design goals and the qualitative character of many factors relevant in the design. Design of mobile service systems requires a balanced mix of hard systems engineering to deal with the technological aspects of the design and soft, collaborative, participative or consensus-oriented approaches to manage the actors in the value network. Therefore, the CBE approach combines the hard and soft system thinking from system engineering. Hard system thinking is useful to deal with the engineering aspects: designing the network infrastructure, mobile middleware and applications. A hard systems approach cannot cope well with situations in which goals are not specified and ambiguous, such as is the case in situations with many stakeholders and in situations where many factors cannot be quantified, such as is the case in services design. Soft systems thinking is more focused on learning processes and ways of seeking consensus among the actors within the value network (Jackson, 1991).

The design activities for solving CBE problems are problem or opportunity formulation, problem or opportunity conceptualisation, model specification, model checking, solution finding and solution implementation (Den Hengst & De Vreede, 2004). The CBE approach puts a strong emphasis on understanding of the current situation before going into a more solution oriented phase. The CBE approach is therefore usually based on incremental improvement, as opposed to radical change. When designing mobile services we think that incremental improvement alone can be dangerous because organisations are locked in their existing processes and may not see mobile service systems opportunities. Relying only on radical thinking, however, is also impossible since organisations and markets have their history and institutionalised habits. The CBE approach supports both radical and incremental change (Den Hengst & De Vreede, 2004).

5.5 Modelling and component-based development

Modelling plays an important role in every design method for systems. Formal models are used to describe the system that is being developed in each of the design steps, and can help to store and communicate the current design ideas. Quantitative models such as simulation models (Shannon, 1975; Zeigler, Praehofer, & Kim, 2000) can be used to compare different design options.

As a formal modelling paradigm object-orientation presents itself as extremely suitable for the design of service systems. It enables modelling and communication ideas from the early stages of the design process (use cases) until the implementation (interaction diagrams, etc.). In line with this, component-based development is a promising way of designing the technical architecture of the mobile service systems. We refer to sections 3.2 and 3.3 for the description of this technical architecture, like service oriented architectures.

Modelling will be described in more detail in section 5.5.1. With respect to conceptual modelling object oriented modelling is applied more often nowadays in the CBE approach, next to flow-oriented modelling techniques like SADT. Object-oriented modelling and component-based engineering potentially align very well. We further look into that in section 5.5.2, and elaborate on a specific form of component-based development in section 5.5.3.

5.5.1 Model-based development

Modelling and simulation play a prominent role in service design approaches, especially in an environment where customer value is key and the customer cannot envisage whether he or she is interested in the services. The systems in which new services are introduced are often complex dynamic systems. It is believed that formal and explicit modelling techniques can and should be used to enhance understanding of the system and facilitate communication between stakeholders and analysts.

A distinction can be made between conceptual models, which define the structure of a problem situation, and empirical models, that represent a further specification and quantification of the problem situation and facilitate analysis and diagnosis of the problem, and that can be used to prototype and research possible solutions. Within empirical modelling, discrete event simulation, visualisation and computer supported gaming are relevant modelling approaches that can also be very useful for mobile service systems design. Sometimes it may seem to be time consuming to grraphically and visually model interactive features of new service systems, but these techniques have a high added value as a vehicle for communication and for gaining insight. In addition, simulation models of services can be used for researching scalability of the architecture, and for further quantitative analysis of the service system in different scenarios.

In product design, models are used to communicate functional and non-functional characteristics to potential users or to the client, to communicate characteristics of interfaces and components between various design teams and to test the design on essential performance criteria. For these purposes product designers use artist's impressions, technical drawings, scale models, prototypes, calculation models and simulations.

Models are used for similar purposes in the design of mobile service systems. The usability of devices, for example, can be modelled in mock-ups or prototypes. In the development of services on mobile devices, emulators of these devices are available for prototyping applications on a PC. The focus, however, will need to be more on models that can express actions or processes in addition to structural or physical modelling. Story boards can take the place of the artist's impressions.

Simulation models (Banks, 1998) can also play a role in researching system aspects such as scalability and robustness in early design phases, which normally can only be tested with a large user community. Simulated users and simulated content take the place of the majority of the expected users, where some of the users can be real at the same time. This 'mixed-mode' approach enables the service designers to study the perceived effect of the system load on the other users, and it can test scalability issues at the same time. An example of a project where this has been applied successfully was the testing of a new timeslot allocation service for trucks in the Port of Rotterdam (Boer & Verbraeck, 2003). A prototype service was developed by several parties, and tested using a simulated road system and simulated container terminals. Several of the trucking companies could use the prototype application, while their competitors were simulated. This allowed us to test the behaviour and response time of the system under a realistic load (Boer, 2005).

This type of use of simulation models is very close to the concept of management gaming (Mayer & Veeneman, 2002), where the concept of interaction with the simulation model is taken one step further. Serious games, as Mayer and Veeneman (2002) call them, can also be used to study interactions between users of the system and to train stakeholders.

5.5.2 Component-based development

Growing commercial pressure to achieve competitive advantage by developing systems quickly, means that component-based approaches are becoming increasingly popular (Cadle & Yeates, 2001). Many varieties exist for these approaches such as Rapid Application Development (RAD) and Dynamic System Development Method (DSDM). The methods take into account management issues and design issues related to the user. Management issues are conditions relating to the environment, empowerment of staff and fit for business purpose. Important elements for a user-centred design approach are the following.

- Testing is seen as being an integral part of the iterative cycle.
- Products are delivered frequently rather than as a perfect end product.
- Interactive and incremental development is an integral part of the approach.
- Users must be actively involved.

Component-based development offers the most promising attempt to meet the challenges in the rapidly changing domains in that it provides the basis of a new set of techniques supporting the next generation of software-intensive solutions (A. W. Brown, 2000). Components are defined as 'an independently deliverable piece of functionality providing access to its services through interfaces'. A component-based development approach consists of the following activities (A. W. Brown, 2000).

- 'Understand the context', the objective of 'understand the context' is to obtain a clear statement of initial requirements. This can be done by use case modelling and business type modelling.
- 'Define the architecture', the aim is to identify all the components that will be used to build the application, and to identify the dependencies between these components.
- Component architecture modelling; for example UML use case diagrams (see Appendix A). The components architecture modelling of mobile services can be structured in layers: the telecommunication network, the middleware, the applications and the presentation layer.
- Context modelling; the aim is here to understand how the system works within the context of the defined business types.
- Interface modelling; the aim is to obtain a set of candidate interfaces, and to begin to describe the details of those interfaces.
- 'Provision the solution'. In a pure top-down approach, the provisioning of the solution starts when the architecture of the system has been completed. In reality this is an interleaved development process in

which architecture definition guides provisioning decisions, and provisioning experiences are used to inform and refine the architecture.

As we can see in the approach above, modelling is the key for different activities of the component based development. It is important to define proper modelling techniques and notations that satisfy the needs of all stakeholders in a development process. The standard object-oriented modelling language UML is the basis for most of the current component based development methods and approaches. The standard UML offers a number of extension mechanisms that enable the user to extend the UML foundation with new modelling constructs according to his purposes.

In the past few years the software industry has been characterised by rapid and continual change in the domains in which software is applied, and in the underlying technologies they target and use (A. W. Brown, 2000). The domain of mobile services is one of the domains in which software is applied. The idea of components is derived from software-engineering environments (Allen, 2001; Allen & Frost, 1998; A. W. Brown, 2000; Heineman & Councill, 2001; Szyperski, 1998; Wallnau, Hissam, & Seacord, 2002), and builds upon the notion of distributed objects (Fingar, Read, & Stikeleather, 1996; Orfali, Harkey, & Edwards, 1996). Three main reasons for using components are relevant for mobile services: (1) reducing delivery time, (2) making use of best-in-class, and (3) supporting parallel and distributed development. To better achieve these benefits in mobile system design, the use of "building blocks" will be introduced in the next section.

5.5.3 Use of "Building Blocks"

Often, complex service systems are built from scratch on top of different network protocols. In addition, the systems have to interface with a number of legacy systems. As a result, architectures are incompatible, there is a lack of integration with existing hard- and software and re-use of earlier developed parts of the system is almost impossible, which results in high investment costs. Therefore, it is expected to be more profitable to design generic building blocks that can be used over and over in different service systems in time (Verbraeck & Dahanayake, 2001). Depending on the type of building block, it can be used in different service systems, and sometimes the building blocks can be so generic that they can be used for completely different types of systems. The definition of a building block used here is (Verbraeck et al., 2001):

A building block is a self-contained, interoperable, reusable and replaceable unit, encapsulating its internal structure and providing useful services or functionality to its environment through precisely defined interfaces. A building block may be customised in order to match the specific requirements of the environment in which it is 'plugged' or used.

An example of a building block could be a transportation service for goods that are bought through a website. If the transportation service is self-contained, interoperable, reusable and replaceable, it can be classified as a building block. When well designed, the transportation building block could easily be used as a part of other services where transportation is needed. Other examples are building blocks for inventory management of the shop, for payment, and for ordering new goods. Each of the building blocks offer a clear service in the system, and on the system level, they hide their internal structure and communicate through a well-defined interface. An example of a component is for instance a universal connector to one of the databases in the shop: an essential, unique element of the solution of the shop that is a (software) enabler for e.g. the inventory management service and the ordering service.

The relationship between a building block and a component can be summarised as follows: a component is the implementation of (part of) a building block in a software environment. The interface (functionality) of the building block and the component are therefore different presentations of the same thing. However, the building block can also be used in an organisational engineering environment. This clear relationship between a component and a building block – where the latter is clearly the broader notion – is supported by the following comments (Herzum & Sims, 2000, p.155).

"A business component is the software implementation of an autonomous business concept (entity or process). It consists of all the software artefacts necessary to represent, implement, and deploy a given business concept as an autonomous, reusable element of a larger distributed information system. Important characteristics of the business component concept are its strong concepts of software artefact ownership, of black box/white box across the development lifecycle, of architectural viewpoints, of distribution tiers; and of its use as a unit of modularisation, of deployment, of reuse, and of management."

The underlying assumption is that a building block on its own cannot provide the functionality a user requires in the sense that a building block on its own – without domain specific context, and communicating, cooperating or even competing with other building blocks – does not provide the required functionality. A search algorithm is nothing without the data to search; a communication protocol is nothing without the sender, message or receiver; and a routing algorithm is nothing without origins, destinations and transporting means, et cetera. A service system, on the other hand, is *fundamentally* a meaningful whole for a user. That means that a service system is an aggregate of more than one building block, where the building blocks interact with each other in such a way that the functionality the user requests is provided. An example: a user wants to retrieve information from a remote geodetic database. To accomplish this functionality we could define a building block for the remote communication and data interchange, a building block for the authentication of the user, a user-dialog so the user can specify his or her needs, a search function in the database and the database itself, et cetera. Together these building blocks form the desired representation.

In order to provide functionality, service systems are constructed by combining building blocks. More precisely, new – conceptual – building blocks can be formed by *aggregation*, which is combining lower level building blocks into new higher level building blocks. We embrace the object oriented paradigm and we want to be able to use the building blocks repetitively in our service systems or models (Booch, 1991; Jacobson, Ericsson, & Jacobson, 1995). *Instances* of building blocks will be combined to construct service systems or models. The instantiation means that the attributes in the instance of the building block get a value in order to give the building block the exact functionality that the user wants within that service system or model. Possibly, a number of attribute variables have a default value so the user does not have to worry about the values of all attributes of the building block.

Based on this the additional properties – inspired by Orfali et al. (1996)– of a building block can be given:

- *It can have different levels of granularity.* Building blocks can be composed of other building blocks and so form larger building blocks with more functionality. Building blocks have an extension that is dealt with in the former property.
- *It fulfils a clear function.* A building block is logically and physically cohesive, and thus denotes a meaningful structural and/or behavioural chunk of a larger system. It is not just some arbitrary grouping.

- *It has a well-specified interface.* A building block can only be manipulated through its interface. The functionality of the building block is addressed by its interface. In that way it is exposed to the outside world.
- *It exists in the context of a well-defined architecture.* A building block represents a fundamental building block upon which systems can be designed and composed. This definition is recursive: a system at one level of abstraction may simply be a building block at a higher level of abstraction.
- *It is nearly independent of other building blocks.* It rarely stands alone. A given building block collaborates with other building blocks and in so doing assumes a specific architectural context. This architectural context is driven in large part by the implementation we chose.
- *It can be used in unpredictable combinations.* Building blocks can be used in ways that were totally unanticipated by the original developer. Typically, building blocks can be combined with other building blocks based on their interfaces.
- *It is an interoperable object.* A building block can be invoked as an object across address spaces, networks, languages, operating systems, and tools.
- *It may have several states.* A building block has some function, and by performing the function, its state may change from state, e.g. from passive to active.
- *It has limited domain specificity.* A building block is defined as being more or less generic.
- *It is a replaceable part of a system.* A building block is a substitute for any other building block, which realises the same interfaces. This aspect helps during development, where parts of a system can be stubbed, sketched, and then replaced by mature, robust implementations. It also supports the evolution of a system, once deployed by making it possible to upgrade and evolve parts of the system independently.

When designing building blocks, one has to take into account three different perspectives. First, the perspective of the *user*, who is the person that uses the service offered by the service system. The second perspective is from the *administrator*, who is responsible for the maintenance of a service system. The third perspective is of the *developer* of the service system. The perspectives are distinguished because the requirements from each perspective differ. The user is usually not interested in the technical structure of the service system itself. He or she is interested in the contribution of the service system to the function, system or organisation it supports. For the user, the advantages of current ICT developments (Internet, Multimedia, Wireless connections) are important. They should be an enabler to manage more and more complex business systems, processes and organisations, rapidly adapt the way of doing business according to ever-changing demands on the market, and/or lead to answers to questions of how to improve the productivity and quality of business services, using these developments.

The following requirements are defined from a user's perspective:

- *Quality:* the information offered should be as correct, timely, controllable, precise and complete, as the user deems necessary.
- *Composition:* service systems must fit together; only in exceptional circumstances they will function in a stand-alone situation. Service bundling is a term used for this.
- *Speed:* the speed with which the service system performs its function must be adequate to carry out the user's tasks.

The following requirements can be defined from an administrator's perspective:

- *Flexibility:* business processes change very fast, so should their supporting ICT infrastructure. In order to accommodate this fast change, the elements should be easily adaptable.
- *Scalability:* in order to be sure that an ICT infrastructure will remain functional when the supported system is expanding, scalability of the infrastructure – i.e. expansion without relative loss of speed or of control – is required.
- *Maintainability:* an ICT infrastructure has to fulfil its function for a longer period of time. During this period the infrastructure environment in which a part of the ICT infrastructure functions is likely to change. This may require change of other parts as well. Besides it is easier and more effective to maintain parts of the service system than by handling it as a monolithic unit, where changes in one part affect the other parts of the system as well.

The following requirements can be defined from a developer's point of view:

- *Generality:* an infrastructure must not only be applicable in case XYZ, but also in other cases of the type XYZ, possibly even in a case of X-something. Tools that can only be used in a specific domain, for a specific goal, in a specific project, do not have added value within other domains or within other projects. On the other hand, if a solution is too generic, practice often shows it is useless. A reasonable number of case variants and some customisable parts may lead to a 'useful-generic' tool.

This kind of generality of an ICT infrastructure could contribute to the speed of development.

- *Reusability:* in order to reduce development time of service systems, reusability of parts is preferred. Reusability does not mean only multi-times usability of the whole software elements, but also of the ideas, methods and design choices. Reusability can be either "as-is", technical oriented, take an element and plug it in without changes to other solutions or "pattern based", more conceptually oriented, from a framework, which can be used to develop various specific tools for a variety of cases.
- *Modularity:* the need for a modular structure of a service system increases, especially when the complexity of service systems increases. That means that a service system consists of a number of highly independent modules.

In conclusion, the expectation is that the development of libraries of building blocks from which mobile service systems can be assembled will lead to more efficient development of mobile service systems. In addition, it will allow more flexible design to adapt to the ever changing demands from the users and the environment. In this design process choices have to be made to find a balance between requirements like generality versus flexibility, etcetera. We further explore the user perspective in the next section, considering its importance.

5.6 User centred design

The service system has to support the activities of the user and has to be aligned with the goals and information needs of the user. Knowing the (potential) users, and knowing the setting in which they will use the service, supports the development of a successful service. It supports the identification of services the user might consider useful, and functionalities that need to be provided by these services.

Depending on the type of user and his context the service system has to satisfy different requirements. These requirements are implications of the interaction between (1) the user, (2) his/her context, and (3) the service concept. The user can apply the service in only in one setting or in different settings; alone or as part of a community. These aspects are relevant to address in order to formulate useful requirements and are therefore further discussed in more detail below.

5.6.1 User characteristics

In the design of a service two types of user characteristics are relevant: Firstly, general human factors such as cognitive, perceptual and motor capabilities and constraints of people in general are relevant. These characteristics influence the way users observe, interpret, decide and act. Think of the limited cognitive capability of a user for example. Secondly, special and unique characteristics of the intended user population in particular need to be addressed. Characteristics such as their education level, knowledge, skills, and experience (see Mayhew, 1999).

In mobile service systems however, several types of users can be discerned. The service can be provided in private as well as organisational settings: users may for example be tourists or truck drivers, as is described in case descriptions in this book. Some services may even be suitable for both types of users, e.g. a mobile service system that supports the booking of hotels.

It is hard to identify which users might be interested in your service. If you ask people whether they would be interested, they can answer this question without committing themselves to the service. As a result they tend to be overly positive, and your estimation of the number of interested users is likely to be too high. Furthermore, it is often hard to imagine exactly what a service looks like when it is not yet fully ready, especially when new technology will be implemented to provide the service. As a consequence, results of test panels might not always be useful in predicting the success of a service.

5.6.2 Context of use

The functionalities provided by the service (utility of the system), as well as the interface applied (usability) should match the context of use of the user. The user may for example be involved in other tasks and processes while using the service. In that case, the service needs to be aligned with these tasks. A secretary, who is frequently disturbed by incoming phone calls for example, has different service requirements than someone who can fully concentrate on using the service. In addition, using speech input to communicate with the interface is not so suitable in a noisy environment. However, users of mobile services, even one single user, are likely to use the service in different environments and under different circumstances. In most cases, use of the service occurs in varying contexts, some of which are unpredictable. As a result, it is hard to foresee the precise context in which the user, the technology and interaction between them will occur.

The functionalities that are provided by the service have to match the goals and sub-goals of the users. Vice versa, functionality that they do not need should be hidden. One strategy to achieve this is to apply personalisation. Storage of user preferences and characteristics contributes to a service that exactly provides what a user needs to achieve his goals. These goals may however shift, depending on the context the user is in. The service system thus should be flexible and also be able to handle these exceptions.

Although, you might not know all relevant aspects in advance, the occurrence of some environmental aspects can usually be estimated. Gorlenko and Merrik (2003) illuminate a number of substantial aspects that are to be considered in the design of mobile systems:

- Environmental variations e.g. temperature, precipitation and lighting conditions might affect performance of mobile hardware. Batteries are consumed more quickly, processors run more slowly at low temperatures and screens are more difficult to see clearly in bright light
- Noisy and distractive environments make mobile systems hard to fully interact with
- Users move, to some extent, from place to place and mobile systems must be able to follow and be usable on the move
- Competition for attention in multitask mobile settings where users are engaged in other tasks and using a mobile system at the same time.
- Users might need to manipulate other physical objects during interaction with mobile systems.

In these five aspects utility and usability issues highly interrelate. The first aspect for example may result in a non-functional device, as a result the service-product cannot be provided. However, even in a case where only the interface does not function, the result is the same: the service-product cannot be provided. And although the fourth aspect certainly needs to be considered in the analysis of user processes and the fit of the service to these processes; smart design of the interface might resolve some of these attention issues. In this case it is very important to attract the attention of the user; the system may for example apply a combination of audio signals and visual signals, instead of only using visual signals.

The criteria presented here all more or less relate to the *notion of fit*. The technology has to *fit* the user, the task and the context of use. Several researchers have elaborated on this notion of task-technology fit which is also referred to as the Contingency Perspective. See for example Benbasat and Lim (1993) and Zigurs and Buckland (1998). Gebauer and Shaw applied the concept in their study of the success factors and impacts of mobile business applications (Gebauer & Shaw, 2004). They investigated the influence of technology characteristics and task characteristics on the usage of the mobile application that provided the service. Poor technology characteristics inhibited the usage of the application to a great extent. With respect to task support users valued two things most: notification, and support of simple activities like tracking, as opposed to handling more complex processes completely on-line.

5.6.3 Assessment of the user and his/her context

In the ergonomics discipline several methods have been developed that support the analysis of all kind of aspects of the user and the setting the user is in; see for example Stanton and Young (1999), Mayhew (1999), and Andriessen (2003) for an overview. Methods include:

- interviews (unstructured, semi-structured, and structured)
- questionnaires
- surveys
- (cognitive) task analysis methods, (see (Schraagen, Chipman, & Shute, 2000)
- (participative) observation
- participative design (Holtzblatt & Beyer, 1996)

Currently participative design approaches are more popular in software design than approaches where experts design the system based on limited input of the users. Researchers found that studying users from the 'outside' is not enough; users had to be involved in the design process itself (Bannon, 1991; Crabtree, 2003). In participative approaches (intended) users of the system are involved in the design process and in evaluation activities related to this process. As Bannon (1991) describes it: "*users are not simply passive objects that others must study and design for. […] People are, or can become, active agents.*"

The information offered by the service system should match the information needs of the user as closely as possible. The interface should offer all the information needed and leave out irrelevant information. Just as user goals may shift, their information needs may shift as well. As a consequence, different requirements concerning the information provided may evolve. These changes may relate to the information itself, as well as to how it is presented. A navigation support system for example may provide an overview of all traffic-jams on the route, or limit this overview to the first 20 km. The system may also provide an overview of traffic-jams within a radius of 20 km, and in that way support the driver in his orientation on alternative routes (or on the availability of restaurant to have a lunch break). In this example, the goals of the user influence two aspects of his information needs: the content of information and the way it should be presented. Requirements for the content of information are related to user processes, and requirements for the presentation of information are related to the interaction process.

With the complicating dynamics of mobile service systems in mind we propose to follow this trend in designing these service systems. Involving users from the beginning in the design of the user process increases the quality of the design. For, the users are the real experts on their task and context. Human factors experts may still have an important role in participative design processes, although this role is slightly different and demands different skills. For users can not fully understand all the possibilities offered by ICT. They need input in order to be able to imagine what the future might be like, using the system to be developed (Kuutti, 1996). This input can be provided by a technology expert for example; someone who knows the capabilities of the available technology very well. The expert and user may be able to construct the future situation together, using scenarios for example. The role of the design expert transforms into a facilitator of this process.

5.7 Process management and project management

In chapter 4, the value networks and organisational network of mobile service systems have been discussed. Mobile service systems include many organisations. Each of these organisations, including operators, content providers, IT-application providers and hand set providers is important at some point in the development process, as shown in chapter 4. Mobile services need to be developed jointly by operators, content providers and IT application providers. In many cases organisations are open for participation in innovative service development projects. It appears that the main reasons for participation, in the explorative phase in particular, are not profit but the strengthening of the image as a high-tech and innovative firm keeping up with state-of-the art developments and initiatives. Another observation made was that often already existing ties between organisations are preferred over new relationships when starting risky innovative projects. Since various dependent actors are involved in a relatively unknown area it is important to create advantages for each actor.

Process management

Involving relevant organisations, creating win-win situations, managing their involvement and the design process requires a way to deal with decision making in actors networks. Process management techniques can help to involve the relevant actors in the design process. As (Checkland, 1999) argues ‘a systems thinker would know that the process of policy creation and policy content are entirely complementary, the process itself conditioning what might emerge as content. Both need to be thought about together; but this is not yet a familiar concept.’ De Bruijn et al. (2002) elaborated on process management. Actors and their values, language and argumentation are important in a process management approach. The focus is on identification and implementation of changes. The design requirements are more than just involving parties, but include a sense of urgency, openness and integrity, protection of core interests and core values of parties, incentives for continuation, and process type arrangements to facilitate sufficient and validated content.

The actors focus on following agreements, meetings and negotiations. The process architect has to design a decision-making process that complies with rules and agreements that result from negotiations with the involved stakeholders. The process manager then schedules negotiating rounds in accordance with the process design, monitoring to ensure that negotiating rules are upheld, and adjusts the process design with the stakeholders when necessary. Gaming is a tool applied with success to design processes.

Process designers have few tools to test their designs. It is difficult to predict the developments and events of decision making in actor networks. One instrument that is applied with some success for this purpose is gaming-simulations (Mayer & Veeneman, 2002).

Project management

Process management can be positioned versus project management. Project management is the traditional form of management used to control projects, including system design projects. In project management it is assumed that problems and solutions are reasonably stable within certain limits, and that clear goals and targets can be set, a fixed time schedule can be followed, a clear framework and a prefixed end product can be used (Meredith & Mantel, 2003). De Bruijn et al. (2002) argue that this only works in a static world and that this approach is impossible for dynamic activities. They add “there will be dynamic particularly when decisions have to be taken in a network. The various parties hold different views about how a problem and a solution should be defined. As a result, the decision making will always be capricious and unstructured”. For the design of mobile service systems there are nevertheless development stages in which project management is a most effective management approach.

Use of management approaches in designing mobile service systems

Both project and process management is needed in designing mobile service systems:

- Process management is required for committing actors to the development of the project and seeking consensus on common goals and negotiating win-win situations.
- Project management techniques are required for managing the development of the project with respect to time and money constraints. Project organisations are needed to embed interdisciplinary design teams within the network of actors.

5.8 Mobile Service Systems design approach elements

In the above we discussed systems engineering, product design, information systems design, collaborative business engineering, component based development, user centred design, and project and process management as sources of inspiration for formulating a design approach for the design of mobile service systems. None of the above discussed families of design approaches completely cover the needs for designing mobile service systems. A combination of elements of the approaches offers a good start to fulfil the requirements for a design approach. This consists of a general set of rules, conditions and characteristics for such an approach. We will address the underlying worldview and principles of such an approach (way of thinking), the managerial aspects of the approach (design process), features relating to the methodical aspect of the approach (design activities) and features relating to the modelling techniques that are suitable for such an approach (modelling methods).

Way of thinking for designing mobile service systems

A central theme in our way of thinking is systems thinking (see section 5.1). In designing these systems it is necessary to address the three elements as discussed in the first chapters: (a) the service concept based on the customer's needs, (b) the 'hard' enabling information and communication technology or the technical architecture, (c) the 'soft' organisational network with its value drivers and actor interdependencies.

With respect to the *service concept* a number of questions around the future use and future users of the service need to be answered. Important factors to assess or define are a profile of the intended users, their level of experience, and their interest in the new service. Around the service itself, it is important to assess the context of use, the user him-/herself, whether the service is replacing one or more existing services, and what the added value of the new service is compared to what the intended users are using right now. Designers have to make many decisions when designing the service concept starting with pinpointing the target group. To get a good idea about the financial feasibility of the service, an estimation of the frequency of use, the number of users, and the price the users are potential prepared to pay should be estimated. Pilots, or even free connection plans to allow customers to explore services can help companies to find answers to these questions. Depending on the kind of service there are many other issues such as localisation-based or not; privacy sensitivity; availability of suitable handsets owned by the target group; content quality and richness; the non-functional requirements.

With respect to the *technical architecture* the limitations of enabling technology should not be neglected in the decision making process concerning a target group and market launch. Examples of these limitations are limited processing power, battery life and memory of devices and limited bandwidth and latencies of the network. These limitations have to be taken into account when formulating the service process requirements; otherwise a potentially good service concept suffers from a bad start due to technical limitations. In chapter 2, we formulated reliability, responsiveness, user interface, trust and customisation as parameters of the service process quality requirements. The level of quality has to match customer demand, and what has been promised must be technologically feasible. A guideline with regard to the technical architecture for designing mobile service systems is to build only services based on *robust* components and not to promise more than is technologically *feasible.* Three other important requirements are to deal with the *complexity* of technological architectures, the *legacy* systems with which the technology component have to be connected, and often the lack of *standardization* to realise easy interfaces.

With respect to the *organisational network* the design of mobile services should be looked upon as designing a complex system with resource interdependencies and constraints that have to be taken into account when creating the network organisation. An actor network perspective should be adopted and it is important to incorporate dynamics: partners join the network, leave the network and get a changing role when they are involved in take-overs or mergers. Agreeing on entry and exit rules when creating a network organisation (for example with operators, content providers, IT application providers and handset providers) for co-development of services can help to deal with those dynamics is important. Reasons for organisations to participate in a network organisation might differ; for example they might be driven by innovation, learning, market entry options, or reputation building. The reason does not have to be profit driven in first instance. Embedded relational ties, either through earlier joint projects or through personal transitions between firms, heavily influence the choice of partners.

Design activities

Design activities for services cannot be serially executed; a waterfall method is too static to deal with the necessary iteration. The design activities have to be part of an incremental and structured work methods, for example the spiral model as introduced by Boehm (1988). A distinction can be made between the early exploration and the later exploitation phase in the mobile service systems design process. In the exploration phase, mobile service design should include radical thinking with at the same time structured tasks. In the exploitation phase a more evolutionary approach is recommended to ensure the buy-in of the various partners and the customers.

The design has to start by determining customers' wants and needs focusing on delivering value to a target group in a specific context. User requirements are often uncertain, necessitating the involvement of users in the design process. Alam and Perry (2002) list the activities performed by customers in the new service development process. Also, design activities used in established participative design methods are recommended (see e.g., Bannon, 1991; Holtzblatt & Beyer, 1996).

The design activities for the functional design have to be done in iteration with the activities for the technical design. The continuous feedback is necessary to realise a design which is technical feasible as well as functional for the user. This iterative way of working is also needed to cooperate effectively with all the involved stakeholders. The chance for a successful realisation and exploitation of the service will increase when all actors are committed to the designed service system. The actors all together also have to agree on the business for the service system. The business case for each partner can be derived from the services' business model.

Furthermore, testing should be an integral part of an iterative cycle, i.e. technological testing and user tests aimed at offering added value and usability.

Design process

Little is known about how the design process for new services should be managed. One of the reasons is that service development differs from product development because of the involvement of the customers in services which requires a customer-oriented new service process (Alam & Perry, 2002). This means that the customer's input in the service development process has to be managed. Furthermore, the design of new services needs the involvement of different kind of actors and organisations. This implies that there will be an ongoing multi-actor decision-making process in different phases (Stevens & Dimitriadis, 2005). The following is necessary to manage this design process.

- The design process should be managed as a project, although process management elements are also needed to commit participants and create an inspiring environment. Participation of stakeholders and experts is very important. Managing their involvement differs in the exploration and exploitation phase considering the different features of the network organisation. Openness and integrity are important to reach achievement between the different involved stakeholders. Use incentives for continuation such as learning and reputation in the exploration phase, and revenues in the exploitation phase.
- Use agreements, meetings and negotiations aiming for the creation of win-win situations, trust is important in the exploration phase and contracts in the exploitation phase. Reducing uncertainty is possible using established relational ties when creating the value network. From the first moment of intention to cooperate on, a common communication strategy shared by all stakeholders is needed.
- To achieve results the use of a project management approach is needed to manage the interdisciplinary teams in which the different roles with their skills are represented, such as creative content developers, innovative application builders and procedural operators.

Modelling methods

During the design process models help to describe and visualise the service system. These models can be used to support communication between designers and between designers and users.

- The service concept can be modelled with modelling tools from the product design and information system design disciplines. E.g. scenarios, storyboards, artist impressions, prototypes and emulations.
- Object oriented dynamic modelling is recommended for the application designs. Appendix A briefly illustrates the modelling technique called UML – Unified Modelling Language, which is currently the most frequently used modelling technique used for designing information systems such as mobile service systems.

- Conceptual modelling of the organisational network is helpful to get an overview of the network's actors and roles. When situations arise with conflicts of interest using gaming can be considered as a way to model the interaction between persons.

Tools

During the design process we can use many support tools to facilitate the design process. The main reasons for using tools are: dealing with and reducing complexity during the design process, communicating design options with other designers, clients, potential users, and other stakeholders, and documenting the design process and its outcomes. Good tools help designers to follow a clear process while generating important deliverables that form the basis for the next stage in the design process.

Many classical design support tools from other disciplines can be used in service system design, such as case tools (e.g. CAD/CAM), presentation tools (e.g. beamers), simulation and gaming tools (e.g. Arena™), diagramming tools (e.g. Visio), application development tools (mostly supplier based, e.g. SDK from Nokia), usability laboratories (see section 5.6), and Group Support Systems (see chapter 7) for user elicitation and supplier negotiation. Each of the disciplines mentioned in this chapter has its own preferred tools, and designers can use any of these tools for service design. The fact that tools partially overlap is not seen as a huge problem, because a good overview is best provided by describing the service from different angles. Recently, literature appears in which is described which tools are suitable for service design, e.g. (Tukker & Tischner, 2006, p.102) mention "blueprinting for visualising the structure of a service, gap analysis for detecting possible problems and quality function deployment (QFD) for minimising those gaps" as existing tools that play a major role in service design. Simons (2006) introduces tools also based on QFD to design e-services in a multi-channel context.

To conclude

The above inventory lists requirements and ideas for a design approach for mobile service systems. These features form the basis for an illustrative implementation of a design approach for Mobile service systems which is presented in the next chapter.

6 Mobile Service System Design approach

An example of an approach to designing mobile services is provided in this chapter. An overview of the design approach for mobile service systems is shown in Figure 6-1.

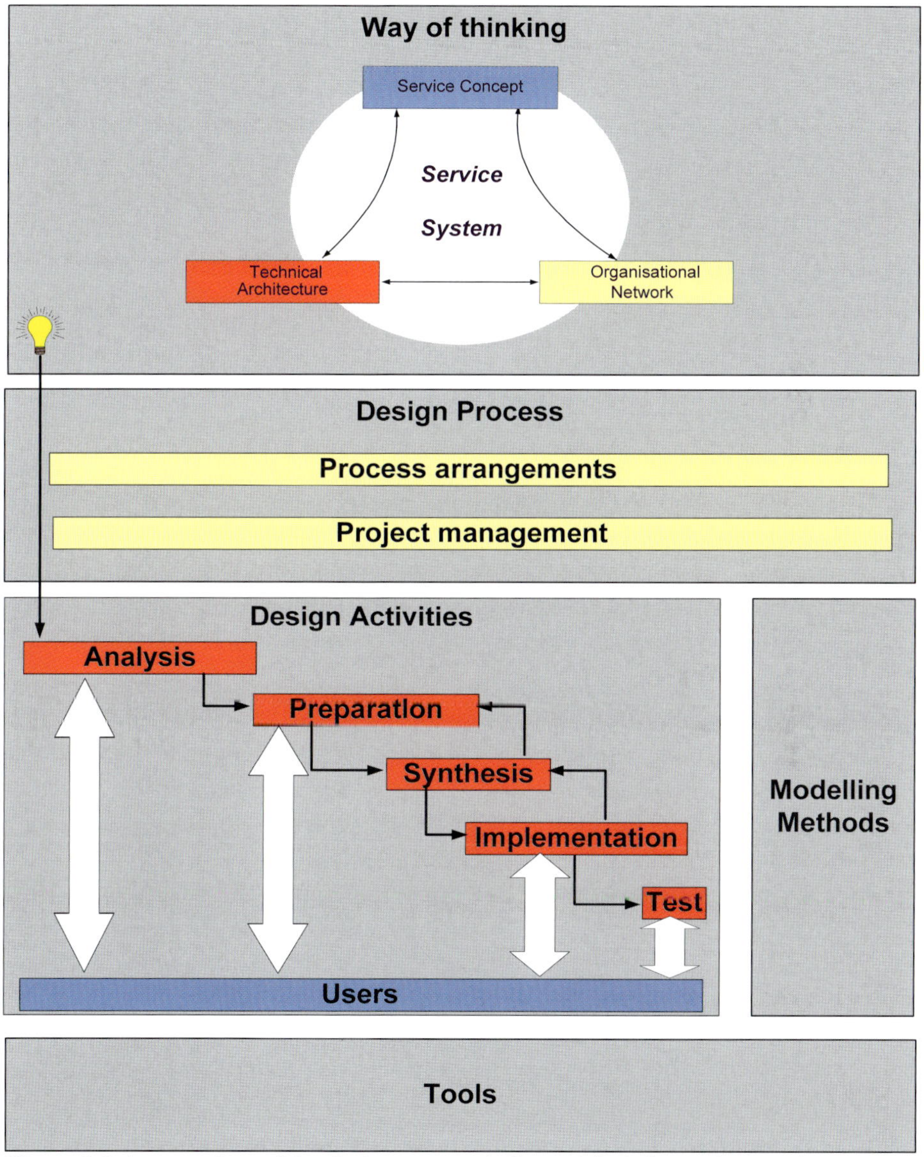

Figure 6-1 An overview of the design approach for mobile information service systems

The approach consists of a way of thinking, a design process, design activities, modelling methods and tools presented in a framework. The way of thinking is used to provide guidelines that can be used as leading principles for the design activities. The design process is the design of the 'design process' which consists of process management and project management activities. The design activities are divided into the phases analysis, preparation, synthesis, implementation and test. Models are used during these design activities. Tools support the whole design approach. The light bulb is used to express an idea that might come from radical thinking and that starts the design process. This design approach will be further explained in the next sections.

6.1 The way of thinking

Mobile Information Services systems result from combining enabling technology, a service concept demanded by users and this service being supplied by organisational network. This model was introduced in chapter 1. In chapters 2 to 4 of this book these 'what' aspects are explained.

The guidelines in the way of thinking are the three trade-offs between the service system elements service concept, technological architecture and organisational network.

6.1.1 Trade-off between service concept – technological architecture

A lot of money has been invested in new technologies in which business developers believed. However, their expectations are often not realised. It is very difficult to know which service concepts users demand based on new technology. The uncertainty in demand causes anxiety levels to rise as it occurs at times when operators are under financial strain. The design approach should include elicitation of user requirements and have user's test prototypes. The market failure of WAP services confirms this. Users cannot be regarded as one big amorphous group. First, different target groups have to be distinguished and representatives of these target groups have to be involved in the design approach. In chapter 2 we concluded that starting the design process with customers' wants and needs in mind and focusing on delivering value to the target group is one of the requirements of our design approach.

Guidelines:

- the targeted user has to be part of the design approach in all phases of the design process; except in the synthesis phase.
- the design approach has to start with the investigation of the targeted user's context, wants and needs.

We argue that the targeted user has to be involved as much as possible during the design process in the exploration and the exploitation phase. However, this does not mean that experts only do what users tell them to do. The experts have to be creative and hopefully come with great ideas and designs. The designers need to have knowledge of the performance of the available technology. Involving the user means that designers can have their ideas tested by users in several phases of the design process. We will elaborate on how to do this in 6.2.

6.1.2 Trade-off between service concept - organisational network

The partners in the value network have to agree on the service concept, the differentiating value proposition offered to the customer. The service concept is determined by dominant actors' decisions along the dimensions of their business models (after Pedersen & Methlie, 2004). Together they have to decide on the value proposition and this can only be based on their perception of the customer's expectations. The gap between the customer's expectations and the service as delivered by the service provider has to be carefully managed (see e.g. (Zeithaml et al., 2006)).

Marketing is crucial for handling the perceived customer value. The company that has a billing relationship with a customer is often considered to be the 'owner of the customer'. This owner takes care of customer relationship management. The operators almost always have the billing relationship in the case of mobile information services. This makes the other actors in the network dependent on the operator. The operator can decide to do the marketing and branding together with the content provider, for example as in some iMode cases where the operator and content providers together advertise in the market [(see e.g. the Radio 538 ringtone iMode service case in (Van de Kar, Maitland, Wehn de Montalvo, & Bouwman, 2003)]. This is different if an independent service provider has the relationship with the customer. A service provider might decide to hide the brand of the operator for the customers. Service providers or content providers can start their own marketing campaign or join with a distributor with a known reputation in the targeted customer market, e.g. a supermarket. Another example is the marketing and distribution of a mobile service packaged with a holiday or conference offer.

6.1.3 Trade-off between technological architecture - organisational network

Fast developing technology is present in all layers involved in producing mobile information services. The layers between the mobile service and the user are the content, servers, content platform area, gateways, networks and clients (Natsuno, 2003). Each of these layers has it's own industry participants and together they produce mobile information services. In the last few years new technology in all the layers enabling new services have came onto the market. Improved technology with regard to the client and the network has had the most impact on service development. On the *client* side new phones have appeared and continue to do so, that are smaller and lighter and have colour displays, built-in cameras, sound chips for playing polyphonic ring tones, wireless connectivity to other devices, and software to connect the device to the Internet. The *network* has evolved, from analogue, to GSM, HSCSD, GPRS, EDGE, and UMTS, and now provides the user with higher transmission speeds and shorter downloading times. To profit fully from the capabilities the client side and network offer, the *gateways*, *content platforms,* and *servers* also have to evolve. If the various firms involved in this process manage to cooperate, the new technology will enable new services that will be adopted by users.

Technology development in the mobile field is quite dynamic. New wireless technology faces competition from other wireless technologies. This means that firms must get services to market quickly to forestall the potential that the technology in which they have invested heavily is bypassed. This pressure is either compounded by or mediated by a firm's general innovation strategy, whether it prefers to develop first-mover advantages or to observe and enter a market later. There are also time pressures related to network rollout requirements that governments place on network operators when granting a license. Although not a direct mandate to cooperate, deadlines for a network rollout often do not allow an operator the time to develop all of the competencies in-house and thus firms are forced to turn to cooperative relationships to procure services. However, the increasing involvement of companies outside the traditional (mobile) telecommunication industry creates longer timeframes for establishing cooperative relations. Thus, despite the increased time-to-market pressure, the development of innovative services across traditional industries takes time. It takes time to establish cooperative relations. Yet time pressure is the case when attempting to get a first-mover advantage in the upcoming UMTS competitive market. The process management elements openness, protection of core values, speed and substance mentioned in the theory of De Bruijn et al. (2002) are helpful here. Incorporating dynamics is required when creating the network.

Guideline:

- take your time to establish a value network and speed up the development process when that is in place.

The need for the flexible creation of a value network puts a heavy demand on the supporting technical systems. New services and processes for service provisioning should be created using a pool of existing and new components and web-services. Openness of the standards used for this is key in being able to couple various services. A radical innovation might originate from a new technology. However, as the service design process starts complexity has to be reduced and the specification should be made explicit (Roozenburg & Eekels, 1995). The service should be assembled from robust components. This will prevent problems such as those that occurred in the first WAP services (Barnes, 2003; Van de Kar, 2002). Technical resource interdependencies are defined by the service requirements; existing technology often does not answer the customer's demands for useful and usable services. The technology interdependencies are extensive; companies have to cooperate to make a service work technically, and such technological interdependencies require a new form of coordination where a proven service IT architecture leads the various applications and content providers in the value network.

Guideline:

- actors in the network can only start to design applications if a proven service IT architecture is provided by the leading actor as basis for the various applications.

6.2 Design process, design activities and modelling methods

The guidelines listed in section 6.1 outline the design process, design activities and modelling methods. These design activities are divided into phases, i.e. Analysis, Preparation, Synthesis, Implementation and Test. In each of these phases we have to deal with the trade-offs between the service system elements.

The design process is controlled by project and process management activities. These activities are all related to the actors and their interests and values. These are covered in design activities in the service element 'organisational network'. The design activities to build the 'hard' part of the service system are related to the technical architecture. All based on continuous input of the users. The deliverables of each design element in each phase are described in Table 6-1.

Table 6-1 Deliverables per phase

	Service Concept (Chapter 2)	Technological Architecture (Chapter 3)	Organisational Network (Chapter 4)
Analysis	Rough service description	Overview technical options	Letters of intention
Preparation	Detailed service description	Functional and technical design	Network creation
Synthesis	Communication formula	Prototype	Work procedures
Implementation	Launched for usage	Operational	Established relationships and processes
Test	Evaluation	Evaluation	Evaluation

We describe the design activities per phase below. The activities are visualised in diagrams for each phase. We number the design activities with first the letter of the phase followed by the letter of the element and at the end the number of the activity in this combination. The duration of the design project depends on the sense of urgency, amount of resources, unexpected incidents and other variables. Models are mentioned if recommended for an activity in the activity description.

The 'APSIT' method described below for mobile information services can be used as a kind of cookbook. Activities can be added in each of the phases depending on the project specific circumstances. It is important that the triad 'organisational network, technical architecture and service concept' is always taken into account. The deliverables mentioned in Table 6-1 are also mentioned in the next subsections in which we further explain the five phases. The exact form of these deliverables depends on the project specific circumstances. It is important to decide at the start of a project what the form and content of the deliverables will be.

6.2.1 Analysis

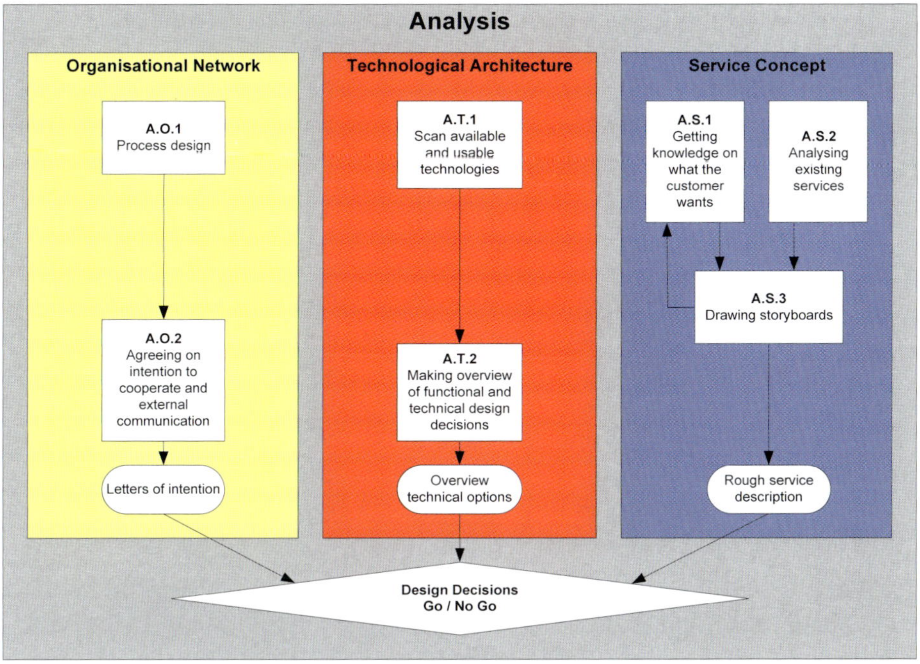

Figure 6-2 Analysis

A.O.1 Process design

First, the process for forming the network of actors has to be run through. It is important to incorporate dynamics in this process and to do this carefully. The creating of the network organisation can be done by executing the following sub activities (based on De Bruijn et al., 2002).

- Exploring the problem; although the process involves several actors, the start of the process will be the concern of one actor or a limited number of actors. If there is a vague idea of the service that will be designed it is time to investigate the available actors that are in the market and willing to participate in the project. The initiator will first discuss the idea with actors that it knows and trusts.
- Actor scans; it takes quite some time to figure out what kinds of expertise will be needed and with what actors 'it clicks'. Since the service provider, operator and content provider will form the core of the network they have to be scanned. In innovation projects the drivers for firms to participate are to support the firm's image, to learn and to achieve a competitive advantage. Therefore it is most fruitfu to approach firms

for which the above count. The table of roles (Table 4-2) can be used as a checklist to figure out whether the necessary roles are fulfilled.

- Quick scan of configurations; there might be conflicting views held by actors that the initiator has to deal with. The initiator can nominate a process architect to deal with this when the initiator has too great a part of the process. The conflicts should be formulated as dilemmas where possible, as framing a conflict as a dilemma has positive effects on the process (De Bruijn et al., 2002).
- Process dilemmas and fixing the rules of the game. The initiator has to make sure that the dilemmas are solvable to an extent that they will not hinder the result. Dilemmas in the exploitation phase might be centred on the revenue model. Is revenue sharing an option if the result is an uncertainty? If the initiator is a large company the situation is different from that where the initiator is a small or medium sized enterprise (SME). In the first case, the initiator might speed up the process by having a governance model in place, as used in portals such as iMode and Vodafone Life. However, this might hinder open cooperation.

When it becomes clear what the network will look like agreements have to be made regarding entry and exit rules, decision-making rules and project management issues such as the organisation of the project, i.e. chairman, steering committee, workgroups, secretariat and the planning and budget. It is recommended that sufficient time is taken to establish the network as this prevents problems later. Modelling methods can help here with actor and role diagrams.

A.O.2 Agreement on intention to cooperate and external communication with main actors

At the point that actors express their intention to participate it is necessary, and important in this early phase, to make clear what information is confidential and what can be made public. It is also necessary to agree on which actor is going to communicate what externally, to the press and so on. One of the issues is to prevent negative publicity regarding first reactions. In innovation projects agreements on confidentially and communication have to be made at the start. This activity is related to the process of formulating the storyboards (A.S.3) and the technical and functional design decisions (A.T.2) since the output of these activities are resource and knowledge.

A.T.1 Scan available and usable technologies

It is a time consuming task to figure out a good combination of handsets, applications and the network. The innovation cycle for each of these differ substantially. One of our guidelines is that the design approach has to start with the targeted user's context. However, since the services are driven by the technology that enables these services, it is necessary to investigate the availability of technologies that are reliable and robust enough to function as an enabler. This also sets the boundaries for how advanced the service will be.

A.T.2 Making an overview of functional and technical design decisions

Between the service and the user is the layered structure of the Internet that consists of the content, servers, content platform, gateways, networks and clients (Natsuno, 2003). The details of this structure will be defined in the preparation and synthesis field. However, based on the output of activity A.T.1, the initiator will have ideas regarding which functionalities and technologies it absolutely does not want to use and which it absolutely does want to use. The output is an overview of remaining options to discuss with the other partners when the value network of actors has been formed. Examples are mobile network technology, i.e. GSM, GPRS, UMTS, WLAN or combinations there-of, positioning technology, i.e. network-based, satellite-based or hybrid and the application protocols, i.e. WAP, iMode or Java (see also chapter 3).

A.S.1 Getting knowledge on what the customer wants

The design approach starts with deciding which target group to address and then with asking the target group about their needs and wants in a certain context. Methods aimed at gathering user requirements can be divided into several categories of which some involve individual users whereas others benefit from the interaction of a group of users. Here we present a user requirement elicitation process to gather useful information in an efficient way early in the design process.

There are three decisions when designing a repeatable process for user requirements elicitation: the participants to invite, the steps to take, and the techniques to use (Den Hengst, Van de Kar, & Appelman, 2004). Each of these three elements will be presented below. Together they form the repeatable process for user requirements elicitation.

Selecting and inviting the participants is the first step for analysing user requirements. The guidelines for selecting participants are:

- Participants need to be reasonable knowledgeable about the topic.
- Participants should be interested in talking about it.
- The group should not include too many different types of people; the participants should share a similar background.

- Participants should be comfortable of talking to each other, but over-familiarity might have a negative effect on the results.

The size of the group depends on the facilitation equipment. Group sessions that are not facilitated by a Group Support System (GSS) ideally have between five and twelve participants. GSS, however, have been found to effectively support large groups (more than 8 members) (Dennis, Valacich, & Nunamaker, 1990). The number of participants is only limited by the facilities used. Facilities are available that can easily host 50 participants.

More than one session is needed when the number of people is larger that the numbers mentioned above. A second reason for having more than one session is the identification of different types of users who do not have a similar background. For each type of user, a separate session should be planned.

Steps to take during a user requirement elicitation session are: Warming-up; Problem analysis; Solution generation; Demonstration of future scenarios (presentation); and the Redefinition of solutions. In this last step the participants start drawing the first designs for the service. These are presented and discussed.

The time schedule for the repeatable process is 3.5 hours in total. Prior experience has shown that three hours is a suitable length for these kinds of sessions (Bruseberg & McDonagh, 2003). The warm-up takes about 30 minutes; the problem analysis is scheduled for 75 minutes. The generation of solutions takes no more than 15 minutes. The demonstration is scheduled for 20 minutes, and the redefinition of solutions takes 55 minutes. With a break in the middle and a closure in the end, the session totals to 3.5 hours. This schedule is based on a group size of 15 participants and the use of a GSS. With more or less participants, the time schedule should be adapted.

Each activity in the user requirement elicitation process requires selecting *effective and efficient techniques* (Bruseberg & McDonagh, 2003). Demonstration of future scenarios can be realised, for example, through a visual presentation or through letting the users handle a prototype. Solution generation can be realised through different brainstorming techniques. ThinkLets can be used for the selection of techniques for the collaborative activities (Briggs, De Vreede, & Nunamaker, 2003). ThinkLets are modular components for building a "script" for the facilitation of the group process. Each thinkLet addresses a specific subtask and is highly structured in order to make it repeatable and easy for a new facilitator to draw on the proven experience on which it is based. The modular nature of the ThinkLets provides for flexibility and adaptiveness in developing the script. There are about fifty thinkLets documented and each helps a group move through one of five activities: diverge, converge, organise, evaluate, and build consensus. Thinklets are often used as part of a Group Support System. For more information on thinkLets we refer to (Briggs et al., 2003).

A.S.2 Analysis of existing services

If the initiator has any idea of the offer she or he has to find out whether the offer is readily available from competitors. The business strategy is different if the mobile service proposition is completely new or has some new service features. The service might also be innovative regarding the underlying technology of the target group in relation to context or the business model. For any of these situations market research is recommended. Reasons for doing market research might include to monitor competitive activities, stay on top of industry events, analyse new business opportunities, or to search out strategic alliance partners (Vassos, 1996). The market research will provide insight into the uniqueness of the mobile service idea and might provide ideas for partners and service features. Furthermore, it is important to take care that the market research and existing service ideas do not hinder lateral thinking.

A.S.3 Drawing storyboards

Professionals add their knowledge, expertise and creativity to develop storylines based on the output of the user requirement elicitation sessions. Modelling techniques originating from product design theories are helpful here. The terms storyboards, storylines and scripts can be used. The result of this activity should provide the first ideas for this and these issues have to be further elaborated in the activities which follow. Usage scenarios can be used to model the service in the context of many interactions. Otherwise, storyboards are sufficient. Photographs, descriptive text and drawn pictures can be used to create storyboards.

6.2.2 Preparation

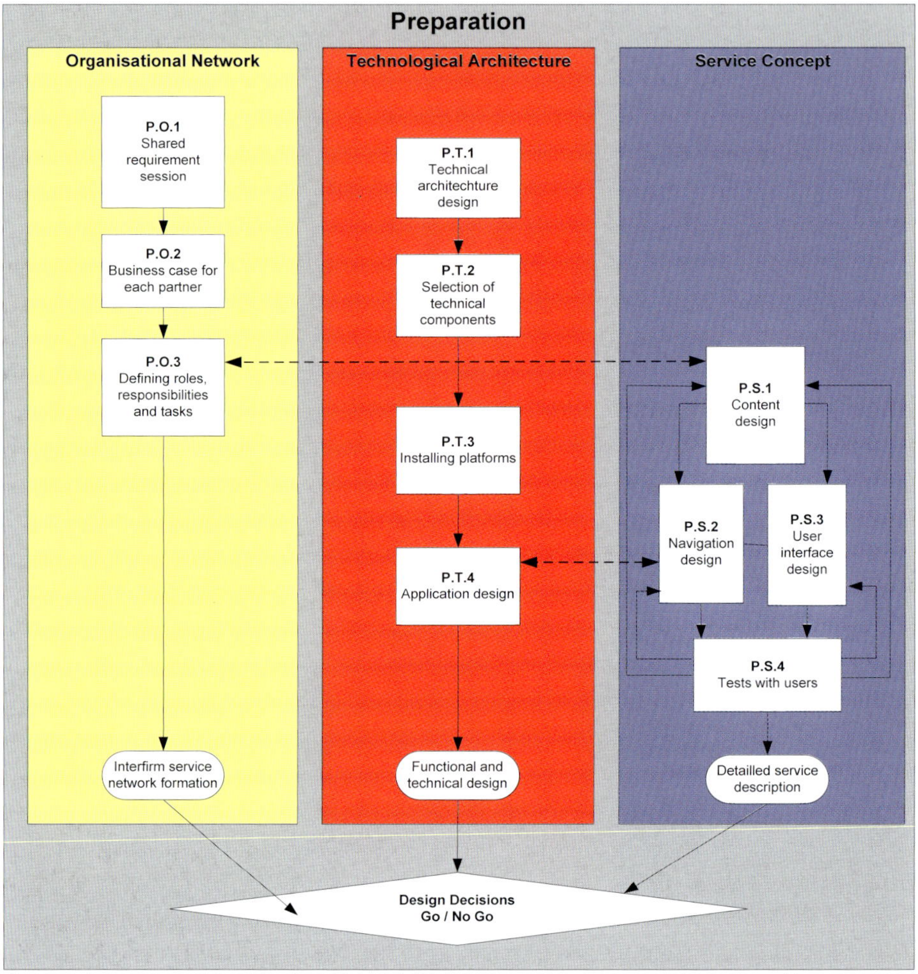

Figure 6-3 Preparation

P.O.1 Shared requirement session

The involved actors have their own expertise, views and interest in participating in the design process. The deliverable of the analysis phase is that they intend to participate in the network organisation. The first activity in the Preparation phase is to agree on the requirements for the service system.

So, once the actors are known an initial meeting can be organised. This can be combined with a collaborative service design session to deal with the complexity of the information service requirements. One of the objects of this should be to come to a shared vision of what the delivered customer value will be. The EasyWinWin methodology with GSS can be used for this. For an overview of the design of such a WinWin approach session we refer to (Boehm, 1988; Briggs & Grünbacher, 2002).

When all actors have a shared view on the value to offer this can be formulated in terms of rough requirements for the service system. The actors involved with the technical design and service concept design can from there on further detail their part of the service system.

P.O.2 Business case for the network and for each partner

During the preparation phase the final decisions are made as to which actors are partners in the project. Every partner has to decide: 'What is in it for us?'; 'What risk are we taking?'. Each partner must develop a kind of business case. In the exploration phase especially the partners' benefits are beyond financial figures. Partners who are enthusiastic about the project have 'to sell' participation within their company; they have to explain the benefits of participation. Based on Afuah and Tucci (2001); Hedman and Kalling (2003) and Timmers (1998) we suggest the following business model components to use as 'hard' evaluation criteria. Be aware that this is NOT on the level of the individual organisation but on the level of the network organisation designing the mobile information service.

- Potential benefits: Are the partners strategically interesting? What is the relative benefit of participating in the value network? Is there an advantage for each partner regarding reputation or learning?
- Cost structure and revenue source models: Is there concordance concerning the division of the needed investment and revenue flows? What is the outcome of a risk analysis?
- Connected activities: Do the partners have complementary resources and skills? Is the partnership good for each partner's reputation?
- Scope: What is the scope of the service development purpose? What is the scope of the market to reach?
- Implementation: Which organisational structure, systems and skills are necessary and available to execute the project?
- Capabilities: What are the capabilities of the value network? What are the shortcomings? Can they be overcome?
- Sustainability: Is it hard to imitate the network organisation? Is there a competitive advantage for the network organisation?

These questions have to be answered for the whole network organisation and based on these starting points the individual actors can define their own business case.

P.O.3 Define roles, responsibilities and tasks

The roles, responsibilities and tasks for the building activities have to be decided in this phase. During the design process specific design roles like application design have to be fulfilled. The actor and role table in chapter 4 can be used to check if the roles are covered. It is important to communicate clearly to all project members which actor is expected to fulfil which role. At the same time co-development of the services between operators, content providers, application developers and handset providers is a must, as shown in chapter 4. It is not recommended to control this process too tightly during an exploration phase. This is a situation with a network structure that has a high density, limited stability and low centrality; therefore the duration should be limited and the frequency of interaction high.

Process management elements are required to guarantee speed in the design process (see also chapter 5 referring to De Bruijn et al., 2002):

- the process should create the prospects of gain and incentives for cooperative behaviour
- the participants should have commitment power
- the environment should enable the process to be speeded up
- conflicts should be transferred to the periphery of the process
- command and control should be used as an incentive to speed up the process.

In this activity project management approaches and tools are also needed to achieve the planned results. E.g. the project management approach 'Six Sigma' and structured product development approach 'QFD', although these methods are very detailed, strict and require discipline (Cohen, 1995; Creveling et al., 2003).

P.T.1 Technical architecture design

The principles of a technical architecture are explained in chapter 3. This can be used to make an overview of the whole technical architecture; this has to be designed from a top-down perspective and then filled in. Jagoe (2003) presents examples of a simple LBS infrastructure and of a J2EE application server that delivers characteristics that are very useful for building mobile location services.

As explained in chapter 5 it is advisable to use a component based development (CBD) approach to model the mobile information service system. The modelling consists of 'component architecture modelling', where a collection of related components are proposed and redefined, 'context modelling', to understand the scope of the system to be developed, and 'interface modelling', in which the interfaces are described in detail. Unified Modelling Language (UML) tools such as use cases, sequence diagrams and component diagram can be used for CBD. See also appendix A.

P.T.2 Selection of technical component suppliers

Once the technical architecture has been constructed the different components have to be chosen. Decisions have to be made as to which components to use and where to get them. What components are in-house and what need to be bought? This has to be discussed and negotiated with possible suppliers. These suppliers might become part of the value network and of the team designing the new mobile information service. Components have to be integrated to deliver mobile information services that take care of the following.

- End-to-end security: Depending on the security required for the service concept it might be a problem if end-to-end security is not available. This is for example a problem in architectures based on WAP.
- Authentication, authorisation and access: one-time password authentication is one of the desired features for mobile phone customers as personalisation further develops (Salonen & Karjalainen, 2003). This is not just for security reasons; it is also to facilitate customers by making the use of mobile services easier. This requires a high level of coordination between the different players in the mobile field.
- Personalisation and profiling; customer profiles are nothing more than a database of information with defined fields (Hagel III & Singer, 1999)[4]. A profile system allows you to personalise services. This often raises concern over privacy abuses. One of the challenges in personalisation and profiling is to provide a simple method for users to tell service providers the information they are willing to share, and under which conditions they are willing to share this information (Jagoe, 2003).
- Positioning: for location-based services, it must be possible to look up and pinpoint a user's location. Different technological solutions exist for positioning. These can be roughly divided between satellite based (e.g. GPS), network based (e.g. Cell-ID) and hybrid solutions (e.g. A-GPS). For indoor solutions technologies like WIFI and Bluetooth can provide positioning information based on the nearest send equipment.

---

4 (Hagel III & Singer, 1999) provide a method for the user profiling process; (Jagoe, 2003) has adapted this for locations based mobile services.

Experiments and pilots with seamless use of those technologies and the usage of it has to provide more insight.

- Billing; the billing system will be related to the chosen pricing model. The current billing systems of operators are mostly designed for post-pay voice billing, quite straight-forward and leave little room for service providers to introduce innovative billing systems. The iMode model is one of the first that gives content providers the opportunity to earn some revenues. Billing systems should be able to handle pre-pay customers, and billing based on number of bytes transferred, URL accessed, application accessed and even quality of service delivered (Jagoe, 2003). The first sophisticated billing products that provide for this are beginning to appear on the market.

P.T.3 Installing platforms

It is necessary to choose platforms on which to run mobile services. Kalakota and Robinson (2002) introduce the following four 'breakthrough platforms' for mobile business.

- Client-side software platform; consisting of operating systems such as Palm, Windows CE and Symbian; and browsers with mark-up languages like WML, cHTML, GML and other XML variants.
- Device platforms; like cell or mobile phones, smart phones, handheld organisers and pagers, specialised industrial handhelds and Pocket PCs like the iPAQ of HP™.
- Hardware platforms; the business-to-business industry with e.g. microprocessor and chipset manufacturers. The major challenge is to standardise the technology. If the mobile service architecture consists of interchangeable hardware and software building blocks it will be able to support a wide range of services across devices, networks and platforms.
- Web services platforms and runtime environments to deploy these applications have to be in place, e.g. JAVA Enterprise Edition and Microsoft .NET.

A number of design decisions on these platforms have to be taken before deciding on an application.

- Many tools can be used during this phase, e.g. J2ME technology consisting of a Java virtual machine and a suite of interfaces that are designed to provide customised run-time environments for embedded and consumer electronics (Jagoe, 2003, p. 115); location interoperability tools from the OMA Location Working Group (www.openmobilealliance.org); end tools from the Platform for Privacy

Preferences, an emerging industry standard promoted by the W3C for automating and giving users more control over the use of personal information at the web sites they visit (www.w3.org/mobile) .

P.T.4 Application design

When the decisions regarding the platforms have been made, several people may design the various applications. Two standards that provide application-level support for wireless networking are mobile IP and Wireless Application Protocol (Stallings, 2002). An application server provides a server-side platform for building and deploying business logic and can provide many technical benefits (Jagoe, 2003). The application design depends heavily on the choices made in P.T.1, P.T.2 and P.T.3. This activity is closely related to the Service Concept activities P.S.1-4.

P.S.1 Content design

This activity is closely related to the application design and navigation design. The content depends on what the users will trigger. The term content should be understood as referring to text, emails, web pages, graphics, audio and video. In this activity the storyboard is further designed to show what the service will look like. This design has to be rich and provide a 'feel-good' impression; the required content is also described in this activity, e.g. the geographical information, databases with restaurant information, games, etc.

QFD is a method that can be used to check whether the content fulfils the customer needs. The planning matrix addresses the following questions for each customer need: "How important is this need to the customer?'; 'How well are we doing in meeting this need today?'; 'How well is the competition doing in meeting this need today?' 'How well do we want to do in meeting this need with the product or service being developed?' and 'If we meet this need well, could we use that fact to help sell the product?'. By filling in the planning matrix including the goal, improvement ratio, sales point and raw weight it is possible to calculate the normalised raw weight with the in QFD provided formulas (Cohen, 1995, pp. 92-122).

P.S.2/P.S.3/P.S.4 Navigation design, User interface design and Tests

A simplified view of the user-centred design process is to conduct user research, set usability goals, design, model and test the user interface; and specify the user interface (Wiklund, 1994). The design, model and test of the user interface should be repeated until the user is satisfied. The navigation design leads to the application design. The user interface design determines the 'look and feel' of the service. The designers develop the font, colour, icons and text. The underlying process of getting a user interface that is usable is a user-centred design process; the targeted user has to be part of the whole design approach. 'Usefulness' and the 'Ease of Use' have to be tested by potential users to determine whether the designers are still working according the users' requirements. This is a repetitive activity. The concept behind it is to stress that customers and technology-providers jointly explore the possibilities and constraints of a technology using simulation (hands-on experience) and developing scenarios. Test sessions consist of a participative evaluation process in which the users actively participate in evaluation and defining usability problems. These aspects can be technical, i.e. a slow device, but also include aspects such as perceived usefulness, i.e. functionality that is not perceived as useful. Different methodologies in usability labs for evaluation may be practiced, including customer field studies, competitive analysis, contextual inquiry, surveys, usability focus groups, participatory design, exploratory lab studies, expert evaluations, and more. To run a usage test you need to organise willing participants, a prototype, a way to make the prototype available to participants, mechanisms for collecting data and a process for interpreting feedback (Isaacs & Walendowski, 2002).

This activity can benefit much from modelling methods, e.g. prototyping to get hands-on experience and emulation to test the user interface. Test and control rooms are handy to use. Interaction design programs like the Adobe programmes Illustrator, Photoshop, Indesign and Flash (for multimedia design) can be used to make user interfaces.

6.2.3 Synthesis

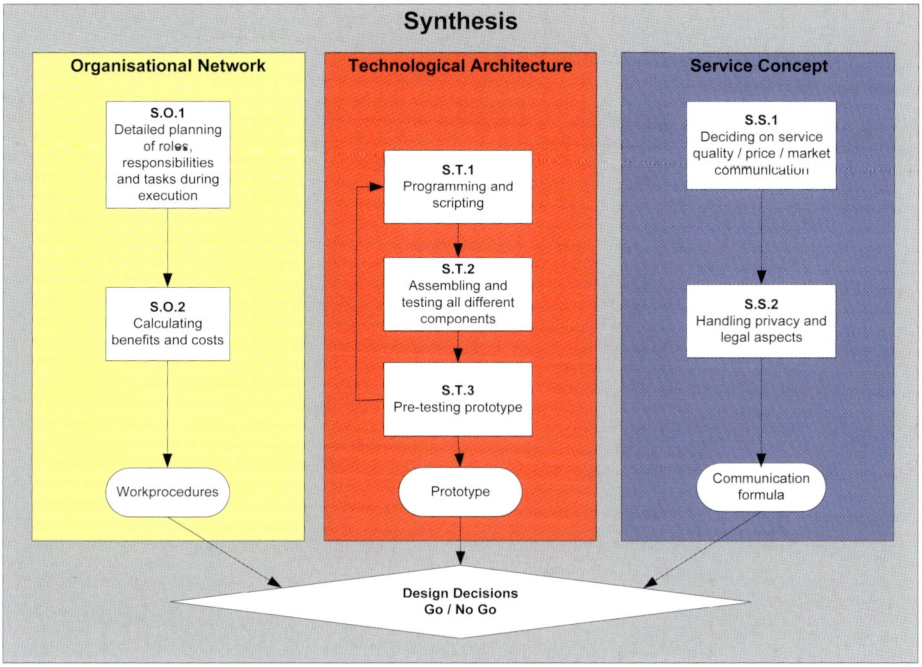

Figure 6-4 Synthesis

S.O.1 Detailed planning of roles, responsibilities and tasks during execution

In the Preparation phase the actors agreed on the role, responsibility and task division during the design project. In the Synthesis phase they have to decide on these issues for the implementation and test phase. This depends heavily on the service life cycle: is this a first prototype test in real life settings, a controlled market introduction for a specified group, or the launch in a region?

The procedure for a first test of the service that will be executed in the next phase has to be completely clear to everybody. This includes the test scenario; it should be clear which researcher does what, when and how they will guarantee that the test results are valid. Workflow models can support this. Practical details like detailed planning for providing and receiving the equipment, a manual for the users, questionnaires, and contracts, the support, etc. also have to be planned. Arrangements have to be made for the operationalisation of the test. What if a server goes down? What if the network goes down? Who is the first line support for the user? If the service will be launched on the market for more than a one-time test period the complete operational processes have to be designed.

S.O.2 Calculating benefits and costs

The results of the preparation phase have to provide an overview of the service system design, while during the synthesis phase, the system has to be built and this costs money. In this phase investments have to be made in the network, in hardware and software. In the benefit and cost model the investment for the next phase, the implementation and test phase, has to be calculated. The risk will vary depending on the size of the test group. The benefits and the costs will be of a financial and a non-financial nature, for example time and reputation.

S.T.1/S.T.2/S.T.3 Programming, scripting, assembling and testing

Many books have been written on developing Internet applications (e.g.: Ince, 2004). These are also applicable in the mobile field. Concerning a value network of various actors that will be located in different places a work method is recommended that is suitable in a distributed environment. A team of programmers in a distributed environment can do the programming if the software is well tested and documented. In the open source development communities, programmers have the adagio 'code somewhere and deploy everywhere'. First the various completed parts have to be tested. Emulators might be helpful here. Next the tested parts have to be assembled and the whole system has to be technically tested. The deliverable of this activity is labelled a 'prototype'. How advanced this prototype is depends on the evolutionary stage of the project. Just before the market launch the prototype should perform in the test as the final service.

To prevent a lot of documentation work afterwards tools like Concurrent Version Systems (CVS) take care of the documentation during the programming work.

S.S.1 Decide on service quality and how to manage the market communication

Grönroos (2001) argues that better quality leads to higher customer retention rates and more profit. The perceived value of the total service offered to the customer has to be calculated during the design process. The developers can calculate the revenue benefit, cost benefit and investment per feature. The further the service is matured the more detailed the calculations will be. Decisions have to be made regarding communication with the users. This is crucial for a market launch, and also for prototype test decisions which have to be made regarding attracting users and how to communicate with them. How will the service be branded, priced, etc.? The whole process of communication with the (prototype) users has to be managed and this can be modelled using the Relationship Dialogue Process Model (Grönroos, 2001). The gaps in the GAP analysis model to manage service quality can be checked to figure out where and how to influence the quality of the delivered service, (Grönroos, 2001; Tukker & Tischner, 2006; Zeithaml et al., 2006 and chapter 2).

S.S.2 Handling privacy and legal aspects

Before the implementation starts it is necessary to check whether all legal aspects are covered. These range from issues like 'can you ask a test person to be personally liable for a piece of technology if it is stolen?' to 'is the mobile information service completely legal?' As mentioned in chapter 2, privacy legislation is an issue for mobile information services that needs to be carefully checked, especially for location-based services. The privacy of the participants has to be taken care of by making sure that they understand what data you will collect and how the information will be used (Isaacs & Walendowski, 2002). For example the Consent Provider Model can be used to manage privacy (Ali Eldin & Wagenaar, 2004).

6.2.4 Implementation

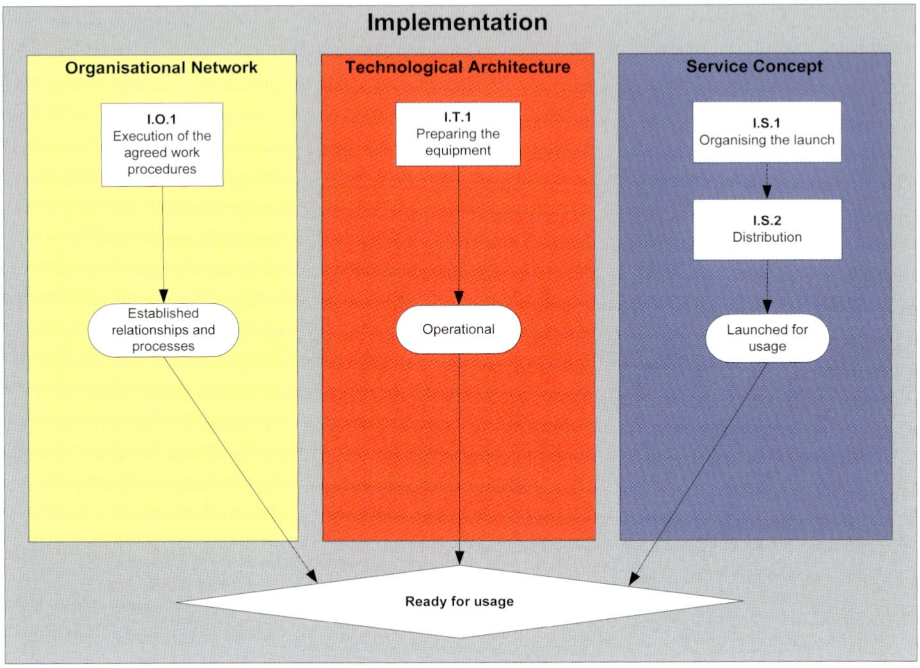

Figure 6-5 Implementation

I.O.1 Execution of agreed work procedures

The implementation and test plan as mentioned in S.O.1 is executed in this activity.

I.T.1 Preparing the equipment

Depending on the number of users in the test group this might be a rather labour intensive activity. The devices have to be pre-installed; a manual with instructions also has to be written.

I.S.1 Organising the launch

The launch might concern a first version or a ready-for-market version of the service to the customers. The contact with the customers should be carefully managed. Results of former tests form the basis for addressing target groups and triggering values of the service. Publicity can be organised when all involved actors are confident with the outcome of the test users' experiences.

I.S.2 Distribution

The relationship with the customers starts once the service is distributed. In test situations the test subjects will receive the equipment, receive instructions and have to sign some form of consent regarding participation. This activity is related to the marketing of the service. One of the issues, for example, is whether the service will be distributed as part of a service bundle. This issue becomes more important closer to a mass-market launch.

6.2.5 Test

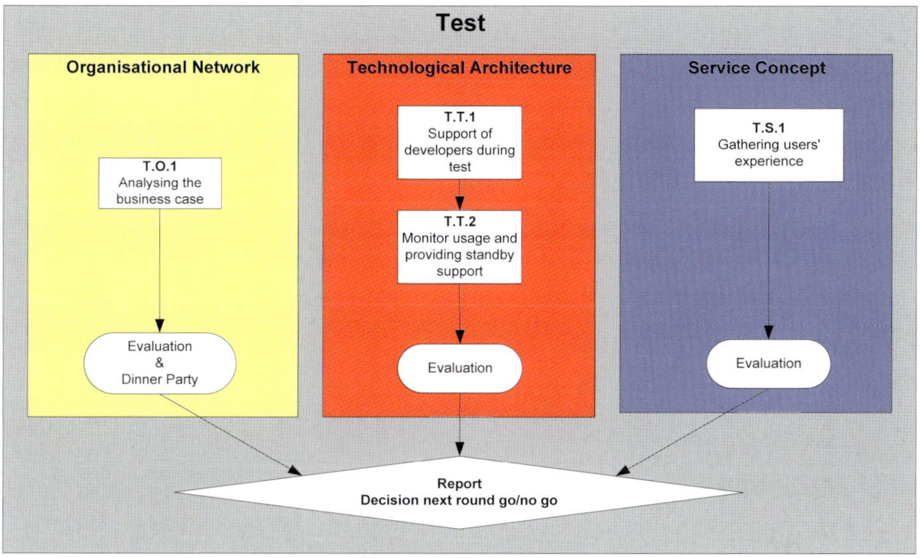

Figure 6-6. Testing

T.O.1 Analysing the business case

The business partners then evaluate their benefits from participating in the design project. They decide if they are willing to participate in the next round. They might use predefined criteria with a 'hard' and 'soft' character to estimate the chances for success. The parameters formulated in P.O.2 can be validated once again. At this stage the partners can also better judge the more 'soft' characterised criteria, as mentioned in P.O.3, and answer questions like 'Was there a good atmosphere?' And 'Are the partners reliable?'

T.T.1 Support by developers during test

The application developers can sometimes solve problems that pop up during the test. This is also an opportunity to have informal conversations with the users and get a better understanding of the usage.

T.T.2 Monitor usage and evaluation

The usage of the network, platform and applications has to be logged. The operators can test the load and peak load on the network. The service providers can use the log files to scale the necessary capacity on the servers, if necessary after running simulation experiments with the logging results as the basis for the empirical models. At the same time, the usage designers can use these log files to evaluate the usage, see T.S.2. Observation is also important for evaluation. Questions that are important to answer for the evaluation of the technology are:

- Is the technology reliable, robust and scalable?
- In what phase of the innovation life cycle is the deployed technology? In other words, Is there an expectation that the technology will become obsolete in the near future?

T.S.1 Gathering user experiences

Data collection begins at the first moment the users start to use the service. The use of a combination of multiple sources to get good feedback on the usefulness and usability of the service is recommended. Questions that need to be answered are: 'what are the users willing to pay for the service?', 'is the service evaluated as useful and easy-to-use by the targeted users?' and: 'is it clear who the target group is and what the size of this target group is?' As mentioned, lessons from case studies, pilots and free connection plans allow customers to explore the services and help companies to find answers to these questions.

There are many methods and tools that can be used to gather user experiences. Examples are the planning matrix of QFD, (see P.S.1.), logging user activity, formal observation and videotaping, informal observation and conversation, surveys. (Isaacs & Walendowski, 2002, pp. 243-273) also provide a good overview for this.

6.3 Next phases

The design approach can be reiterated after a first round has been completed and a test group has used the service. In the second round some activities may be modified and might be done more quickly. After each round the service can be distributed towards a larger test group. After two or three rounds the object of the approach will be the market launch and the start of exploitation. In the last round before exploitation more emphasis should be placed on the business model of the service and the business cases for the various partners. Exploitation might start with only a small part of the target group. After the first real market experience more market exposure risk can be taken. The number of necessary rounds will vary per situation.

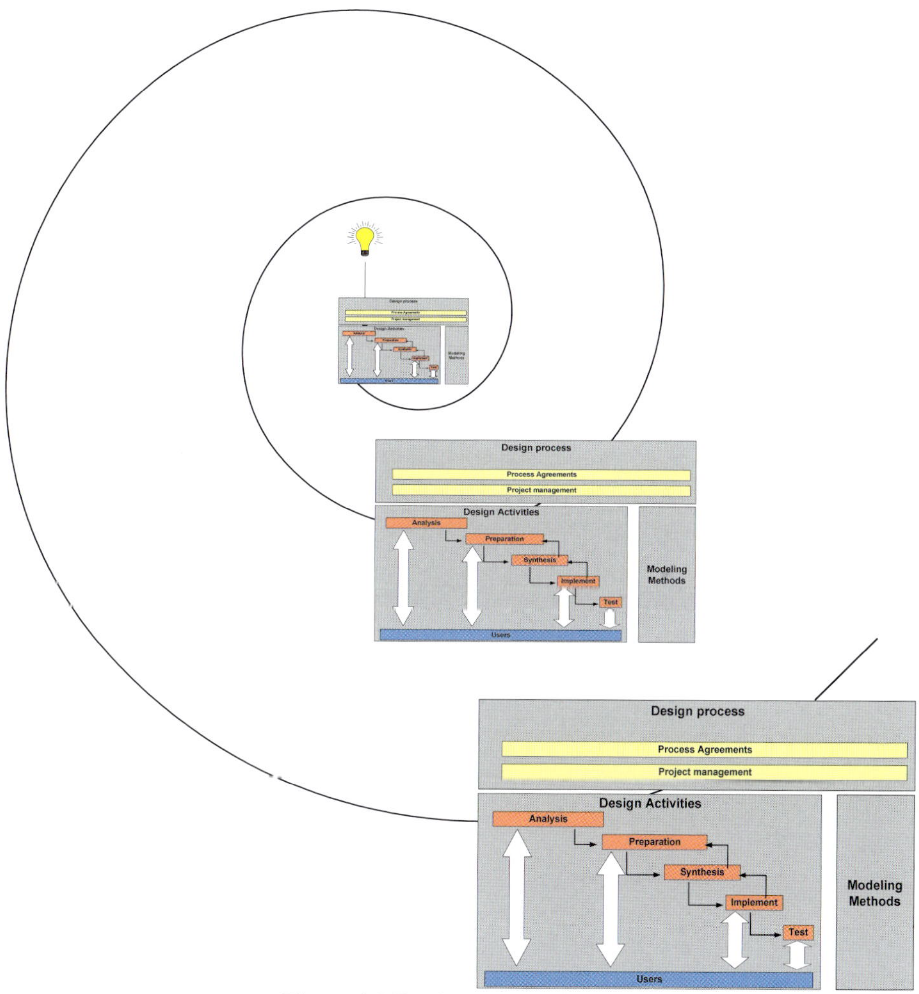

Figure 6-7 Design approach spiral

The whole design approach spiral is presented in Figure 6-7. The light bulb is used to denote that the whole process has to start with radical and lateral thinking. Iterative design steps should be taken within each round; and after each round follows an incremental step towards the next round.

The described design approach can be modified and extended in all directions: technical details, usability aspects, distribution, marketing, social interaction, etceteras, etceteras. Designing service systems is a multidisciplinary and complex process. The approach described in this chapter can be used as kind of checklist. In next chapter the experience of applying this design approach in a first round in an action research case is described.

7 Case: MIES on the campus

This case describes the experience of applying the design approach explained in the previous chapter. The occasion was that Delft University of Technology and the mobile operator T-Mobile organised a UMTS test-bed at Delft, in the Netherlands. A mobile information service system for campus visitors, called 'MIES on the campus', was designed in one of the projects on this test-bed. The service system consists of a service concept, technological architecture and organisational network. The service concept was based on storylines derived from brainstorm sessions with users. The available technical building blocks were a UMTS network in a test-bed setting, servers, a UMTS phone; a PDA including a navigation application; and a GPS receiver. The organisational network consisted of a mobile operator, an IT firm, content providers, game developer, a navigation software supplier and developers from the university. Together they designed, built, implemented and tested 'MIES on the campus'.

7.1 Background

The Faculty of Technology, Policy and Management (TPM) within Delft University of Technology (TU Delft) is concerned with complex technological systems which function within a socio-economic context (www.tbm.tudelft.nl). The Faculty develops interdisciplinary theories, methods and tools to address technologically complex problems. One of the research themes is concerned with the design and evaluation of innovative service concepts; the focus being on services which have been made possible by ICT and that are designed in a multi-actor context. The mobile service field is one of the domains in which the researchers apply their research. These researchers were interested in opportunities for cooperating with industry to apply and test the design methods and tools they had developed.

In particular, the mobile telecommunications industry is interesting since it is searching for new services, not only to regain its investments in licenses but also to stay competitive in the future. The industry is undergoing a radical transformation and organisations in the mobile telecommunications, information technology and media sector are becoming increasingly interdependent. Organisations have to collaborate in complex networks to create new mobile information services.

These first years of the 2nd millennium, mobile operators are working hard on their plans to roll out the third generation mobile network according to the European standard: Universal Mobile Telecommunication System (UMTS). In 2002, mobile telecommunication operators started to make plans for pilots to investigate all user and technology issues related to UMTS. This system provides broadband mobile connectivity Europe-wide. In 2003, manufacturers were delivering the first network systems and terminals were appearing slowly on the market. However, the new services that had to be delivered, were - and still are - in the development stage. One of the five Dutch mobile operators, T-Mobile, approached Delft University of Technology to exploit a UMTS test bed at the university.

The UMTS test-bed initiative allowed researchers to design an advanced mobile service system. A test-bed can deliver valuable information about the users experience of intended services. This also concerns the new terminals that can be used with UMTS. A university campus environment with students and staff members can deliver important inputs for the further design of the services and terminals. The campus environment can be seen as a breeding place for ideas for new services generated by the creativity of campus users.

One of the working packages on the test-bed was the project '*MIES on the campus*'. MIES is an acronym for Mobile Information and Entertainment Services. The service '*MIES on the campus*' used the features as offered by UMTS, for T-Mobile the purpose of the UMTS test-bed was to show the public the advantages of UMTS. For the university, the focus of the project was on concept exploration, prototyping and testing of innovative services on this new broadband wireless infrastructure UMTS. The specific purpose of the '*MIES on the campus*' project was to apply a developed design approach that focused on the value network issues: how to create a value network of actors and how to coordinate the activities between those actors.

The challenges of the project were to develop and test a mobile information service system that fulfilled the following criteria.

- To provide a mobile information service that is useful and usable. For mobile services this is based on needs and the service must be usable in a setting where the situation of the user is critical concerning location and timing (Anckar & D'Incau, 2002; Baldi & Pyu-Pyu Thaung, 2002).
- To provide an innovative service. The innovativeness of a service can either be related to the technology used, target group in relation to context, the business model used or new service features/applications.

- To provide a service system that makes use of existing building blocks as much as possible. This meant that the application should consist of available building blocks with only some programming to assemble the building blocks. The reason for this is that the research should be focused on service design and not on the design of technical components.
- The last criterion concerns the multi actor environment. The service system had to be developed in a value network with various actors who fulfilled the roles of the mobile operator, service provider, application developer, content developer(s), hard and software suppliers etcetera. This situation reflects the reality and fits into the long term multi-actor research programme within the Faculty of TPM.
- Furthermore, the aim of the project was to develop a service in a relatively short period of time. Arrangements for the UMTS network occurred during the autumn of 2002 and the spring of 2003 and the network was operational in June of 2003. The time available to develop the *'MIES on the campus'* service on the network was less than five months (May, June, September, October in 2003; hardly any work was done in the holiday months July and August). The operation of the test-bed network benefited from the experience of the T-Mobile staff with their own test-bed in their local office.

The *'MIES on the campus'* project started in April 2003 and the project was divided into five phases: analysis (April-August 2003), preparation (August-October 2003), synthesis (October 2003), implementation & test (October-December 2003). The outcome of the design activities for the organisational network, technical architecture and service concept respectively are described below.

7.2 Analysis phase

In the analysis phase the rough outline of the value network of actors became apparent, there will be an overview of the technical options and there would be an idea of the service concept.

7.2.1 Analysis phase - Organisational network

Below the backgrounds of the most important partners in this network are further described. Founded in 1842, *Delft University of Technology* is the oldest, largest, and most comprehensive technical university in The Netherlands. With over 13,000 students and 2,100 scientists (including 200 professors), it is an establishment of both national importance and significant international standing (www.tudelft.nl). The Faculty of Technology, Policy and Management is the youngest faculty of the university. The focus of the research and education of the Faculty of TPM are problems with technical aspects as well as organisational, administrative, economic or legal aspects (www.tbm.tudelft.nl). The faculty cooperates closely with business and government organisations in research centres and research projects to solve social and corporate problems. The TPM researchers involved in this research project are part of a research programme in which theories, methods, and tools to support design and evaluation of (distributed) organisations, processes and policy in the services sector are being developed. Furthermore, two other faculties were involved, one to provide knowledge on product design, usability, and ergonomics and another for their knowledge on GIS and GPS.

T-Mobile Netherlands provides mobile telecommunication solutions to 2 million active customers in the Netherlands (January 2004). T-Mobile International is the shareholder (100% subsidiary of Deutsch Telekom). T-Mobile and the other branches of Deutsche Telekom serve a large number of customers worldwide, about 80 million (www.t-mobile.nl). T-Mobile in the Netherlands had a special group of people assigned to prepare the UMTS launch.

Another important partner in the project was *Delft Plaza*, this is a local content provider who maintains a website with all local information on events, shops, restaurants, etcetera (www.delftplaza.nl; in Dutch).

Ex Machina is an independent, Amsterdam, Netherlands based developer of games. Ex Machina has enabled successful multiplayer games for Web and Mobile since 1998, with millions of games played. All their work is based on their unique multiplayer technology Deus+ (www.exmachina.nl). Another application and content provider in the project was *AtoBe*. This firm makes web enabled applications available on all current and future mobile devices, browsers and technologies (www.atobe.nl). Their role was focussed on video applications in the project.

The Vision Web is a consulting firm with about 400 consultants and delivers ICT solutions to their customers. A group of consultants specialised in IT architecture was involved in this project (www.thevisionweb.com).

A set of standard roles related to mobile services was used to search for actors to build MIES (see chapter 4). Embedded relational ties were without doubt a very important factor in the network formation. Everyone involved in the project knew at least on other person involved in the project, as a teacher, (former) colleague, business partner or personally. T-Mobile was already involved because they had already worked on the initiative of UMTS test-bed at the campus with colleagues from Delft University of Technology. T-Mobile was also the connection to the handset suppliers and video application provider. A content provider offering local content on Delft and a multi-user game developer were found via the personal relationships of researchers at the TPM faculty. The IT organisation was a former employer of the researcher. Expertise of researchers of other faculties at the large Delft University of Technology was determined via personal relationships. Only TomTom was chosen for experience with their product, at the start of the project nobody knew anybody in the company personally. This influenced the cooperation with them, there was only contact a few times by phone and email and they did not really participate in the design project. For the test-bed project more companies were contacted than those that actually participated. The reasons chose not to participate with more companies were: not enough time to get familiar with the technology, nobody had experience of the company and they asked too much money, or simply nobody got in touch with them.

The network was recognisable as an exploration network with high trust/openness and low control (Nooteboom & Gilsing, 2004). None of the actors asked for a letter of intention during the network creation phase.

7.2.2 Analysis phase - Technological architecture

T-Mobile provided the conditions and determined what technologies could be used regarding the network and the handset. As a pre-commercial implementation, the test-bed network represents a particular incarnation of the UMTS technology and as such presents a platform with unique features and functions.

The UMTS network test-bed was built to be partially integrated into the existing T-Mobile GSM network. The way this network has been set up means that no phone calls can be made to the GSM network from the test-bed in Delft (and also not to the fixed PSTN network as this is also linked to the GSM network) but data transfer is possible to and from the Internet, and GPRS could be used, provided the mobile phone supports an appropriate browser. As the WAP gateway is also located within the GSM network, WAP services also cannot be used.

It was a difficult task to figure out what the best technological solution was for localisation. The choice was made for GPS since Cell ID was not available. Another decision that had to be made was the choice between a client or server based application. A big advantage for a server-based solution is that the information can always be updated, e.g. information on roadworks provided as and when necessary. A disadvantage is that the user always has to go online via the mobile network, which costs money, can cost extra time and has a chance of failure.

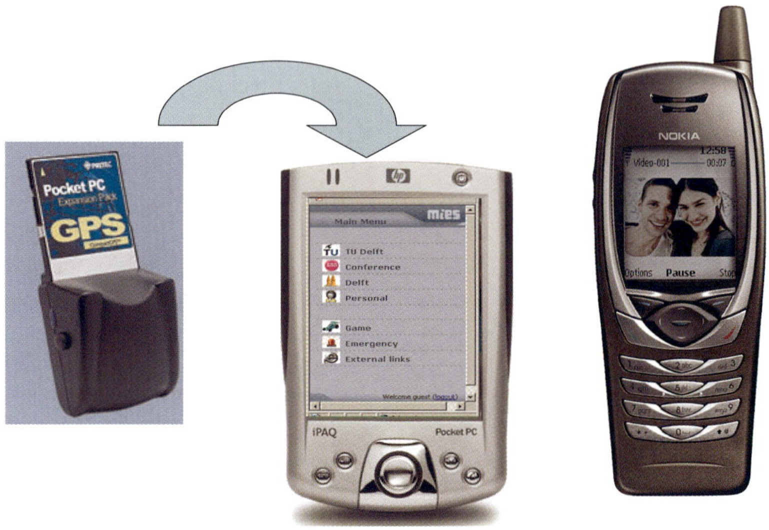

Figure 7-1. Chosen handsets

It was a time consuming task to figure out a good combination of handsets and the UMTS network. The starting point was to determine what was available on the market. E.g. a UMTS phone with integrated GPS was not available. The UMTS phone Nokia 6650, the iPAQ2210 and a GPS receiver were the devices chosen to enable the service. The connection between the Nokia and the iPAQ was via Bluetooth. The GPS receiver was a CompactFlash card that fits the iPAQ, see Figure 7-1.

7.2.3 Analysis phase - Service Concept

One of the dilemma's in the 'service concept – technology' trade-off is that there is no known demand profile for services based on new technologies like UMTS; there is no history of use and no customer experience base and reliable survey data to draw on. You also cannot ask customers what they want if they have no idea of what you are talking about. One way to solve this problem is to start with the context of the user.

The *'MIES on the campus'* project started with brainstorm sessions supported by Group Support Systems (GSS) with visitors to the campus to find out about problems they encountered when they visited Delft. Three different groups of participants, being foreign academics, Dutch academics and Dutch practitioners were defined and a GSS session was held with each group. In a structured way the target groups provided information on the problems they encountered when visiting the campus (for a detailed description of these sessions see Den Hengst et al., 2004). The result of the selection of the most important problems is shown in Table 7-1.

Table 7-1 Contexts for which a solution (service) needs to be found

Session 1 **Foreign academics**	**Session 2** **Dutch academics**	**Session 3** **Dutch practitioners**
Local Logistics by public transport (when, where, how)	Do I have to walk or is there a bus or a taxi and how expensive is a taxi (personal up to date route planner)?	Details on the appointment (location, time, attendees)
How can I get in touch with my fellow countrymen?	Restaurant facilities	Information on other people around the campus
Where can I find information about educational persons?	Information on courses (person, time, place, teacher)	What eating and drinking facilities are there?
Where do I get emergency help and medical care if I need it?	What facilities can I use?	What educational, scientifical, and social events are taking place?
What's the weather forecast?	Can I find a person with the same interests and meet this person?	
Where can I find information about cultural/ recreation/sports events?	What events are going on where?	

Based on the results of the sessions storylines were developed as input for the service descriptions.

Seven storylines were formulated based on the outcome of the GSS sessions. To make the storylines realistic a situation was based on a fictitious person. This description was as follows.

> *'A researcher, 'Mabel Jones' from Great Britain visits the conference on simulation from 26-29 October 2003 in Delft. Mabel arrives on the 25th of October at hotel Dish in Delft and receives a package at the reception desk with a UMTS device, iPAQ and GPS receiver that together provide the 'MIES on the campus' package.'*

Storyline 1: find the conference venue

The next day, Mabel reads the instructions and starts using MIES to work out where she is, where the conference location is and how to get there. She could walk, take a taxi or go by public transport. The answer is that it is a 10 minutes walk. So she starts walking using the handheld device to help her find the route. She finds her way easily.

Storyline 2: find the right session

Mabel arrives at the venue and registers. She can choose between three parallel sessions that start in 15 minutes. Two of them are relevant for her. Mabel uses MIES to download the abstracts of the papers from the two relevant sessions and she makes a choice. She decides to plan the rest of the sessions she wants to attend later. During the morning session she receives an update of the programme. One session is cancelled; another one delayed and there is another keynote speaker.

Storyline 3: find the right people

The morning sessions are over and lunch starts. Mabel wants to find some people who are involved in the same research issues as she is. She starts MIES 'getting in touch' module and searches for people based on key words she enters in the search fields. She receives pictures and descriptions of people that match her request. She recognises Juan Gonzales from Spain and sends him a message inviting him to sit next to her at lunch. The lunch is held at the conference venue and Juan and Mabel have an interesting talk during lunch. They exchange business cards via their devices.

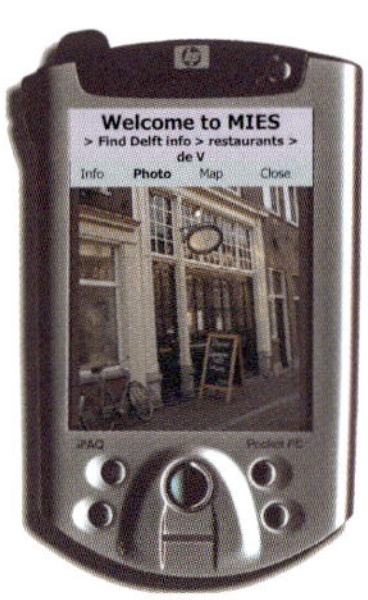

Storyline 4: find a nice restaurant

The first day there is no dinner organised by the conference committee. Mabel and some other conference visitors decide to go into Delft to have dinner. They search with the help of MIES for a nice restaurant. They want an informal restaurant with a reasonably priced menu in the centre of Delft. They can choose between several places, one of them has a webcam and they can have a look at the actual situation at that time in that place. They decide to go there and call the restaurant to make a reservation. MIES advises them to take the bus to the restaurant and MIES provides them with the bus number, where and at what time the bus departs and how to get to the bus stop.

Storyline 5: tour through the campus

The second day in the afternoon there is no session that Mabel and Juan like and they decide to explore the TU Delft campus. The MIES service has a tour that guides them through the campus and that provides interesting information on the buildings, e.g. about the library and faculty buildings; where famous professors work and their research fields; the current courses available at the university; and they find they can download a preview of a lecture from a professor they know. Practical information such as where the coffee machines are, copiers, ATM's and lunch facilities is also available.

Storyline 6: more time to kill

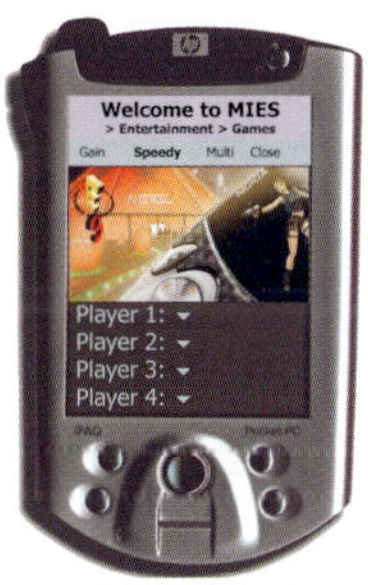

They end the tour and still have some time to kill before the official conference gala dinner. They decide to play some multi user games on the mobile. Then they receive a message that there is an extra social meeting at Speakers, a big pub/dance venue downtown, which they can attend along with people from the conference. They take a look at the situation via a webcam in Speakers, and they see what appears to be a good party. They decide to go there and get directions via MIES.

Storyline 7: appointment info

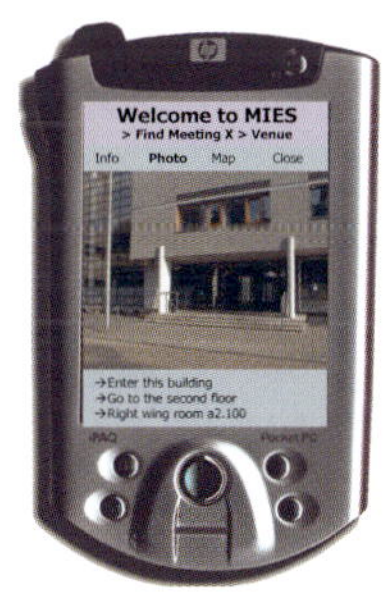

Next day the conference is over but Mabel has an appointment with an assistant professor at the Faculty of TPM. The appointment is at 09:00 o'clock. Mabel is late because she slept through the alarm (due to the party the night before?). She phones a taxi and sends a message to her appointment that she is late. In the taxi she has a quick look at the profile of the person she will visit. Thanks to MIES she can easily find the room in the TPM building where the meeting is. She recognises the person from having seen her picture on MIES.

7.3 Preparation phase

In the preparation phase the organisational network becomes established, the functional and technical design will be made and there will be a detailed service description.

7.3.1 Preparation phase - Organisational network

In this phase the final value network creation was finished, the result is shown in Figure 7-2.

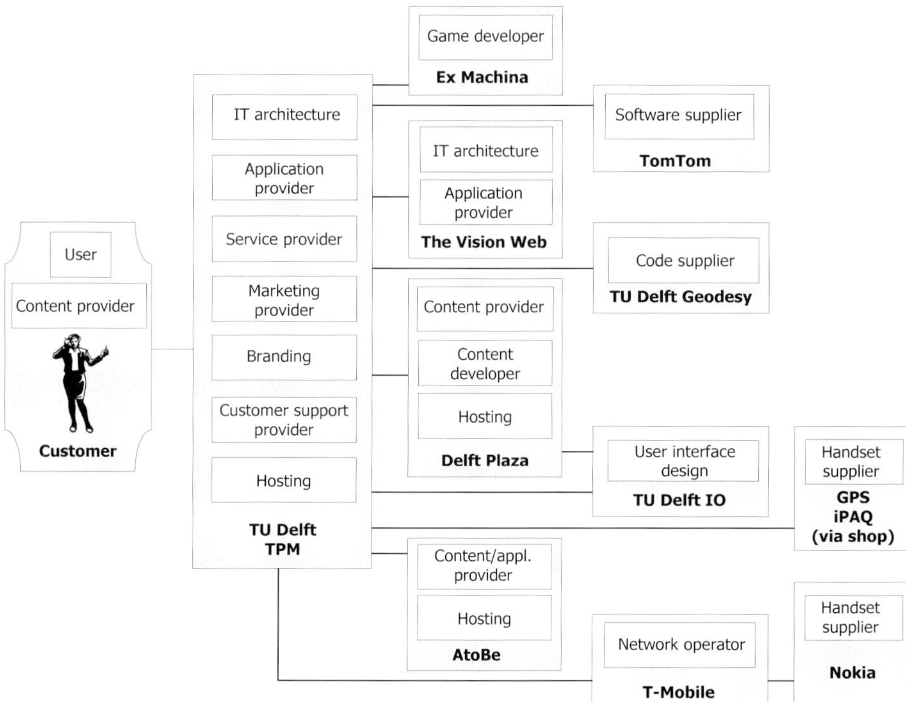

Figure 7-2. MIES value network

At the start of this phase the project initiator of the university organised a launch meeting, including drinks afterwards, with all involved actors. They told their background and plans and hopes for the project. Researchers of the Faculty of TPM presented the first ideas for solutions. As shown in Figure 7-2 many roles were provided by this organisation. A consultant from The Vision Web, an IT architect, was appointed technical project leader. He organised a meeting every week and made minutes providing action points for all these meetings. Two application programmers of The Vision Web later contributed to the programming work that had to be done. Ex Machina had developed a multi-user game on GPRS and wanted to test this for UMTS. They hired a student to do this as a study project. Delft Plaza is the local content provider in Delft and maintains a website for inhabitants and tourists. Their business model is that they are paid by shops etc. for the companies to be mentioned on the Internet. They considered offering their content on the Mobile Internet to be a good extension of their business. This project was a good experiment to start to gain experience. T-Mobile offered the UMTS network in the overall UMTS test-bed project of which '*MIES on the campus*' was a sub-project. In this sub-project T-Mobile's role mainly dealt with choosing and arranging for availability of handsets and keeping the network 'in the air' during the experiments. T-Mobile was also the actor who brought in AtoBe to test video applications in the project since video applications are suitable for testing the specific features of UMTS such as speed. The navigation programme of TomTom was first just bought at the shop but when it became clear that licenses were necessary for all the iPAQs the firm was approached and asked whether they could sponsor these software packages, and they did. The faculty of industrial design participated in the navigation and user interface design. Two students worked on this as a study project under the supervision of an associate professor. They collaborated intensively with programmers of Delft Plaza who had to convert their Internet content to the mobile Internet. Geodesy researchers were involved in providing knowledge on GIS and GPS conversion.

More detailed information on which actors fulfilled which roles in the MIES case is listed in Table 7-2.

Table 7-2 Provided roles by actors

Role category	Providing Role	Definition	Explanation for developing	Organisation
Functionality related roles	Service provider	Provides billable service to the end consumer	Service development	TU Delft TPM * Service will not be billable.
	Network operator	Operates the mobile telecommunications network over which the data (service) is transmitted	Designing and building the network	T-Mobile
	Platform provider	Provides the software that defines the general platform on which a variety of services are run	Preparing servers	Delft Plaza TU Delft TPM

Role category	**Providing Role**	**Definition**	**Explanation for developing**	**Organisation**
	Application provider	Provides the software that makes a service possible and that sits on top of the platform	Application development: personalised applications location based applications game web application	TVW and TU Delft TPM TU Delft TPM Ex Machina Delft Plaza and TU Delft TPM
	Web hosting/presence provider	Operates and maintains the server that hosts a website that is an integral part of the service, particularly relevant to the further development of content	MIES website hosting MIES finding people hosting (ASP) MIES finding people hosting (PHP) Game hosting	Delft Plaza TU Delft TPM (server witlab) Server at TU Delft TPM (Section ICT) Ex Machina
Content supply chain roles	Raw content supplier	Supplies content, but format is not usable yet for the mobile service & terminal	Supplying content for MIES website Supplying content for Game Supplying Location information	TU Delft Marketing and Communication Users Municipality, house of chambers (raw content providers behind Delft Plaza) BluTarsky Game Players TomTom

Role category	Providing Role	Definition	Explanation for developing	Organisation
	Content developer	Transforms raw content into content appropriate for the service as well as the mobile terminal	Developing MIES website	Delft Plaza
			Developing personal information sites	TU Delft, TPM
			Developing Game	Ex Machina
			Filtering existing information and make it suitable for MIES	TU Delft, Industrial Design
	Content provider	Provides 'appropriate'/trans-formed content to the service provider	Providing MIES website	Delft Plaza
			Providing user data in a way that it fits in MIES	TU Delft, TPM
			Providing Game	Ex Machina
			Providing video clips	AtoBe
	Content aggregator	Serves as an intermediary between the service provider and the content providers	Navigation design User interface design	TU Delft Industrial Design
Hardware roles	Equipment provider	Provides the hardware (physical components of network)	For the UMTS Network	Via T-Mobile
			Server	TU Delft, TPM (witlab)
			Server	Delft Plaza
			Server	Ex Machina
			Server	TU Delft, TPM (ICT section)

Role category	Providing Role	Definition	Explanation for developing	Organisation
	Handset supplier	Supplies platform or service-specific handsets		Nokia 6650: via T-Mobile iPAQ 2210: paid by T-Mobile; bought at shop GPS receiver; paid by TUD; bought by shop
Customer relation roles	Billing provider	Provides billing services to collect revenues from end consumers		Not applicable
	Marketing provider	Markets the service		Not applicable
			Agreement on external communication	TU Delft, TPM and T-Mobile
	Customer support provider	Point of contact for customer queries regarding the service; responds to customer queries	During the test Ex Machina and TU Delft, Faculty IO were on standby to answer questions	TU Delft TPM
	Content quality manager	Monitors and improves content quality		TU Delft TPM Delft Plaza
LBS roles	Content geocoding provider	Adds x/y coordinates to the content		Delft Plaza
	User positioning	Provides the position information of the mobile device		Via GPS + MPU
	Positioning technology vendor	Supplies user positioning equipment		Not applicable; done by GPS
	GIS provider	Provides geographical information (and GIS services) necessary to indicate location information of relevant content		TomTom Delft Plaza

7.3.2 Preparation phase - Technical Architecture

The functions of MIES are ordered in a Presentation-Logic-Data architecture, see Figure 7-3. This overview was made as a start for the technical design of the MIES service. The presentation layer is on the client side (plain); the data is on the server site (dots) with the exception of the TomTom maps that are on the client site. Satellites provide the GPS data and these data are converted by a script to GIS coordinates. This is further explained below regarding localisation. The logic layer (stripes) describes the actions between the client and the server. Again, the logic that is only related to TomTom is completely on the client site (plain). The data layer was extended with content from other content providers who joined the MIES portal later, e.g. the video clips of AtoBe. This layer can be extended indefinitely.

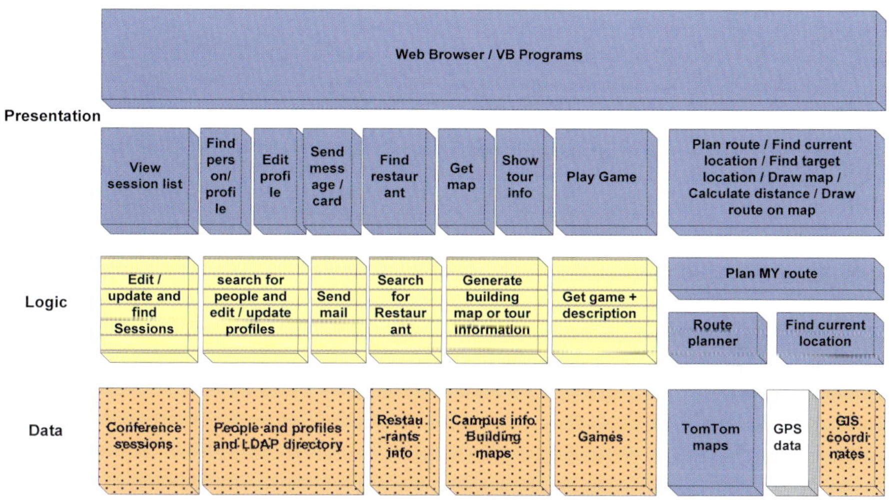

Figure 7-3. Presentation-Logic-Data architecture

Besides the Presentation-Logic-Data architecture a Technical Architecture was also made, see Figure 7-4. This technical Architecture consists of the following building blocks.

- The UMTS network: a 3G broadband packet-based transmission of text, digitised voice, video and multimedia at data rates up to 2 Megabits/second theoretically; in practice the maximum is much lower.

- The Internet: The Internet fulfilled a central role in MIES: all information to be retrieved through the MIES portal was requested and provided via the Internet.
- Content: Multiple content providers made their content available for the MIES portal via a connection from their private network (via a firewall) to the Internet. Of course, this required the right configuration of various network equipment, e.g. routers, firewall, at the content provider's network and the telecom provider's network.

The MIES technical architecture is flexible: new content providers can be added to the web portal, they can be "clicked into the value network" easily.

MIES enabled users to find another person (finding people) and to find buildings (finding a restaurant). For both of these functions the users had to know what their current position was and receive the requested location. A client-based solution was chosen for the location-based function to MIES users. There were more standard components available for this and the TomTom Navigator, a standard navigation package for iPAQ users, was chosen. The maps are all stored locally on the iPAQ.

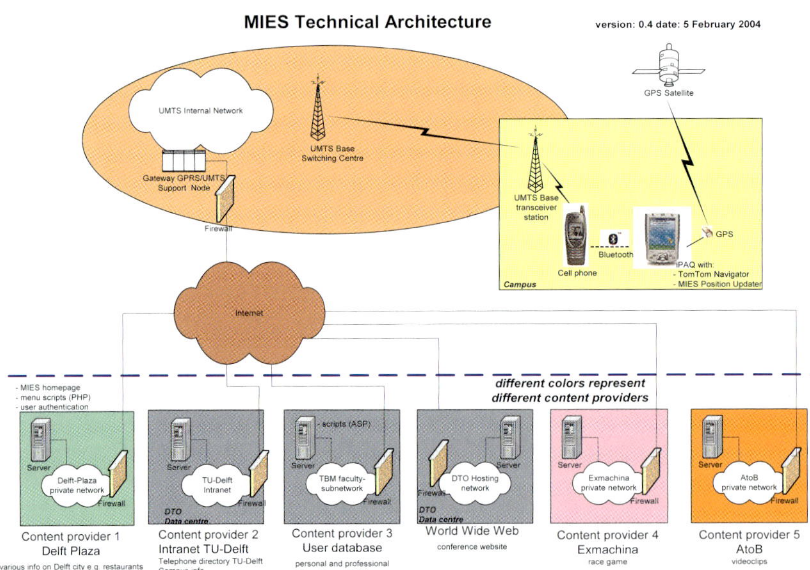

Figure 7-4: Technical architecture

7.3.3 Preparation phase - Service Concept

Based on the storylines and the available raw content, user interface designers made the navigation structure and user interface for MIES. Websites for tourists, e.g. of London and Rotterdam, were also checked. The design was made in a few iterations, where users provided feedback between results. Two user test sessions with three test persons each session took place. The first session was a user studio session with three test persons taking around 30 minutes for each test. This was the first test to get feedback on the navigation structure after two weeks of designing. The pictures in Figure 7-5 illustrate the project's usage studio session.

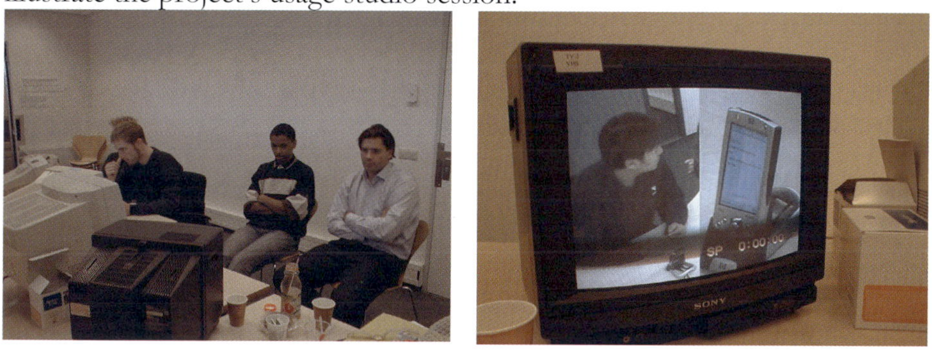

Figure 7-5. Control-room and Test-room

A few weeks after the studio sessions the second user test session, a field test with three different test persons, was held. For this test all content was available and the test persons walked outside so that the designers could observe *'MIES on the campus'* being used in the field.

7.4 Synthesis phase

In the synthesis phase the service system is actual built. The focus of the activities is the technical architecture. The involved actors prepare themselves for the implementation phase by defining activities, roles and preparing decisions on the marketing aspects of the service concept. Users are not involved in this phase of the design process.

7.4.1 Synthesis phase - Organisational network

During the synthesis phase of the project, the interdependency of the technology and the value network became very clear. Some of the programmers wanted to program in Java, some in PHP and some in ASP. In the end some actors programmed a part of MIES in PHP and other actors another part of MIES in ASP. This was not the preferred choice, however considering the available resources this was the best available in the allocated time span. The consequence of this was that two servers were needed; a UNIX and a Windows environment, and two similar templates had to be made. In the second version of MIES for the student test it was decided to program everything in PHP. This meant that the ASP partner left the project.

Ex Machina worked stand-alone. This was not a problem for the service since the game was a client side solution on the Nokia but this situation was not so nice for the programmer. This was a student who wanted to learn from the project and as such was hindered since there was not much collaboration with him.

People worked under time pressure in this synthesis phase and it was very hectic trying to get the designed service ready in time. For example, T-mobile ordered 20 iPAQs but somehow the order did not come through. One day before the implementation was supposed to start project members drove to a shop with enough stock and bought the iPAQs. None of the actors bothered about calculating benefits and costs because everybody's focus was on realising the service.

7.4.2 Synthesis phase - Technical Architecture

Various programmers from different organisations worked on the programming of the functionalities. The functionalities related to finding people and finding buildings were supported by the Mobile Position Updater (MPU). For this, a PhD student formulated requirements for the following functionalities: user authentication, look and feel management, separation of the business logic from the look and feel by means of templates and MySQL database connection.

The MPU is a programme written in Embedded Visual Basic that runs on an iPAQ pocket PC. The programme name is "MIES Position Updater.vb". The following software had to be installed on the iPAQ: the eVB Runtime, TomTom activeX control and the programme file to run it. The MPU fulfilled two functions: (1) send the current GPS location of the iPAQ to a web server, and receive a (latitude, longitude) position from the web server in return and (2) when the web server returned a position the TomTom Navigator map-view had to show the (latitude, longitude) location on the screen. The setting screen and the information flow between the iPAQ and the web server for finding a person are shown in Figure 7-6.

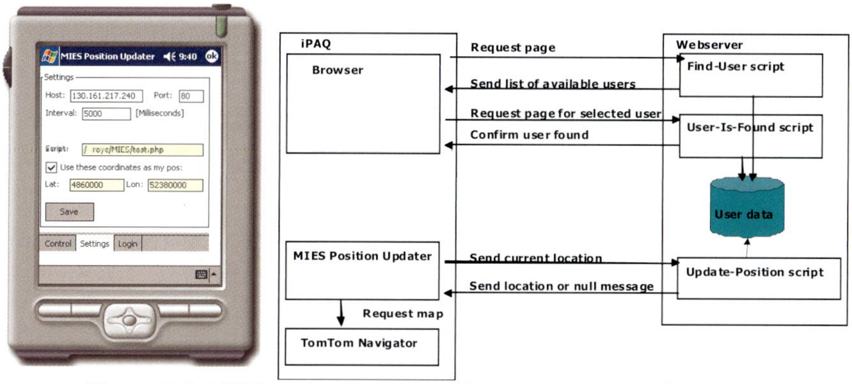

Figure 7-6. MPU setting screen and locating persons information flows

The same principle as shown in Figure 9-6 was used for locating buildings. The difference was that the information on these buildings is held in a database. The database of Delft Plaza is the information on restaurants including the XY coordinates according to the GIS standard. This GIS information had to be converted to the GPS standards longitudes and latitudes so that the MIES Position Updater could request the map in the TomTom navigators.

The PHP scripting was implemented in the following WebPages 'Login screen', 'Main page', 'Finding other user's locations' and 'Chat'. The vAuthenticate system was downloaded and used for user authentication. This is a ready-to-use user management system that supports a number of levels. Furthermore it comes with some code to check user permissions for every page that has to be protected. User data on the client side is stored in cookies. These PHP scripts were assembled using the PHP scripting of Delft Plaza that made their content available for the iPAQ. The Vision Web worked on an ASP application for the personal settings. This was a standalone solution that also was stored on another server. The multi-user game Razor was programmed standalone for the client side, the Nokia.

7.4.3 Synthesis phase - Service Concept

The following service applications were developed based on the storylines and using the available building blocks and resources.

Goal 1: Finding locations. This functionality was supported by a navigation programme, called TomTom, that runs on the iPAQ. The GPS knows the location of the user and the user can insert the address he or she wants to go to and TomTom will then show the route. The locations relevant to the conference visitors were pre-installed as favourites in the TomTom programme. Since GPS does not work within buildings it is not possible to guide users once inside buildings.

Goal 2: Update of the conference programme. There was up-to-date information about the conference programme provided via the Internet on the *'MIES on the campus'* PDA pages. There was also a function to alert people to important information.

Goal 3: Finding people who 'match'. People could search for other people who share the same interest, research field or speak the same language. In the first prototype of the service, users were able to search for other subscribed users only, while in the second prototype, they could search for other subscribed users and send messages. A privacy control functionality was embedded in this service (see Ali Eldin & Wagenaar, 2004).

Goal 4: Finding a nice restaurant. If you are in a strange environment you need help to find restaurants, shops, hotels etc.. A selection of information that was already in the database with local content of an Internet content provider, and all that was relevant for conference visitors, was translated into English and formatted into the right format for the *'MIES on the campus'* PDA pages; including search functions, information and pictures of the restaurants and the locations. The MPU script took care of providing the requested location to the user.

Goal 5: Killing time: playing games. Mobile games are played on mobile phones and not on PDAs. Part of the MIES project was a multi-user game called Razor. It was not interesting for a game developer to invest time in developing a game for the PDA. Thus, Razor ran on the UMTS phone instead of on the iPAQ. Users anywhere on the campus could race against each other. The game was already available on GPRS and was adapted for UMTS. The advantage of UMTS for multi-user games is the improvement of the latency. This is more important than the availability of more bandwidth.

Goal 6: Guided campus tour. The guided campus tour consisted of information about the buildings of TU Delft presented on the *'MIES on the campus'* PDA pages. Part of the information was instructions on how to walk round the campus to view the various buildings. This could also be viewed using the TomTom navigation programme. A Delft city tour was also developed in addition to the campus tour.

Goal 7: Being on time at the right place for an appointment. Address information including phone numbers and room numbers of TU Delft employees were obtainable on the iPAQ via the LDAP server of TU Delft. However, it was not feasible to give direct access to the agenda of the people. Thus this storyline was only partly executed. In Figure 9-2 an example is given for one of the TU Delft paths in which MIES users could find information about TU Delft on people, locations, facilities and transport.

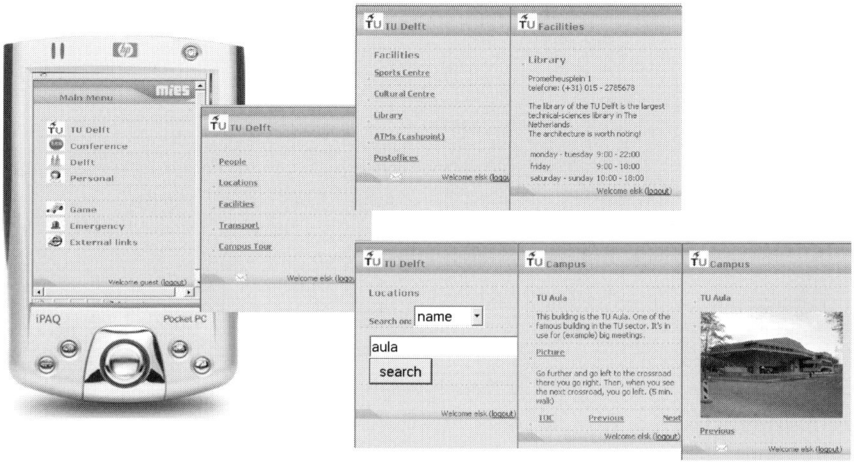

Figure 7-7. 'MIES on the campus' screen dumps – example path of TU Delft submenu

7.5 Implementation and test phase

At the end of the implementation and test phase the results on the design of the organisational network, technical architecture and service concept are evaluated. Measuring tools are important. Implementation and test took place very iteratively and is described together.

7.5.1 Implementation and test phase - Organisational network

In this phase the relationships between the actors were established. The partners evaluated the benefits and costs from participating in the design project. The benefits are mentioned in the test results (section 7.6). The partners did not make the analysis of the business cases separately. Theoretically the initiator did make a start with looking at the revenue model for the service.

7.5.2 Implementation and test phase - Technical Architecture

In the implementation phase, two people dealt with the pre-installing of the test equipment and wrote the manual. This was done interactively so the manual text could be checked for accuracy. This took two people about a week. Then, the applications had to be installed on the iPAQs and the

Nokia's; the three devices had to be connected and tested; all the terminals had to be put in special bags and the manual had to be added. This was a very labour intensive task that had to be done precisely. Four student assistants spent an entire day on this for 18 sets, see Figure 7-8.

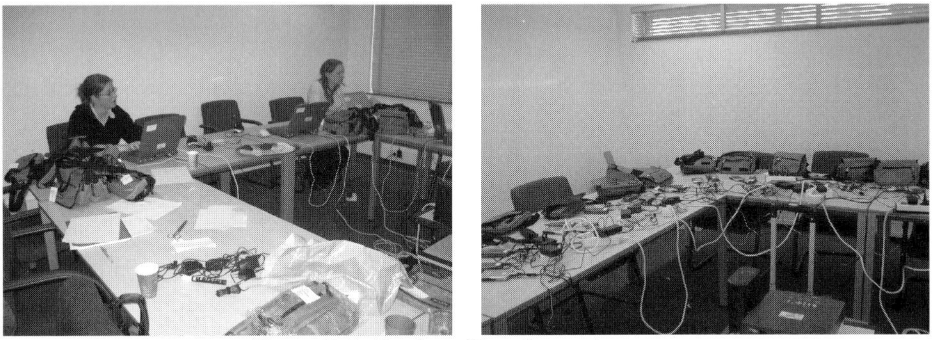

Figure 7-8. Pre-installing the equipment

During the tests there was a first line and a second line help function in place. Sometimes the test persons needed help with the software settings etc.

7.5.3 Implementation and test phase - Service Concept

With regard to the user test the decisions were made to test the service using attendees to a technically oriented conference. The test persons did not have to pay for the service. They only had to sign a contract in which it was stated that they should take good care of the equipment. They would not be liable for any equipment damage that was not the test person's fault.

The prototype test was organised during the 15th European Simulation Symposium (ESS) in Delft and symposium attendees were asked to use and test the MIES service for three days. The ESS conference is a conference on simulation and all the attendees were researchers in this field and can thus be considered to be early technology adapters. This was a good group of people for a prototype test. The 15 test persons came from all over the world, had different ages and were mainly male.

After analysing the results from the first test round the initiator wanted more data. Therefore, a second test round was organised. Master students taking a course on Service System Engineering performed the second test. One of the assignments in this course is to design a service. Testing MIES for a week allowed the students to experience the complexity of putting ideas into practice.

An overview of the first and second tier pages of the MIES menu during the tests during the conference and the course is provided in **Table 7-3**. The results of both user test rounds are described in section 7.6.1.

Table 7-3 Overview MIES menu

ESS conference	**SSE course**
TU Delft	TU Delft
People (LDAP server)	People (LDAP server)
Locations	Locations
Facilities	Facilities
Transport	Transport
Campus Tour	Campus Tour
Conference	SSE course
News	Check News
Schedule	Schedule
Find people	Find people
Locate people	Locate people
	Send messages
Locations	Locations
Transport	Transport
Chat	Chat
Delft	Delft
Find a restaurant/pub	Find a restaurant/pub
Locate restaurant	Locate restaurant
History	History
Walk in Delft	Walk in Delft
Sports	Sports
Other	Other
	Personal
	Update my Profile
	Receive message
Other	Other
Game	Game
Emergency	Emergency
External links (web browsing)	External links (web browsing)
	Video clips
Check email	Check email
Other	Other

7.6 Test results

The findings can be divided into the results of the user tests and the results of the partners' experiences.

7.6.1 User tests results

The extent to which the test persons felt MIES was useful is shown in Table 7-4 on a scale of 1 to 7. Note that '7' means that they strongly agree that it is useful. The first round test persons gave a 5.4 for usefulness; this is good since it is more than the neutral 4.

Table 7-4 'MIES on the campus' service performance results

	First Test (mean; n=15)	**Second Test (mean; n=22)**
Usefulness	5.4	2.3
Clarity	5.1	4.4
Efficiency	4.9	3.8
Support/help	4.8	3.8
Satisfaction	5.2	3.9
Visual qualities	5.0	5.1

Table 7-4 shows that the first round test persons answered all service performance parameters positively, meaning better than the neutral 4. This was not the result of the second round test group. There is a big difference in the results between the groups regarding the usefulness of the service. The fact that the first group was the real target group (conference visitors) and the second group was not (the students) seems to have had a big impact on the usefulness judgment.

Besides the more general service performance, the test persons were also asked to give their opinion on specific MIES functions. The users' opinions on specific MIES functions are shown in Table 7-5.

In the second test round more questions on specific functions were asked. It shows that some items were given a negative answer by the test persons; finding an employee, the Delft City Tour and Chatting all had a mean below 4.

Table 7-5 'MIES on the campus' function results

	First Test (mean; n=15)	**Second Test (mean; n=22)**
Finding buildings	5.2	4.2
Finding people	4.5	4.1
Finding restaurants	4.7	4.8
Campus tour	5.2	3.5
Razor game	4.3	4.6
Finding employees	-	3.5
Finding locations	-	5.1
Delft city tour	-	3.9
Sending messages	-	4.1
Browsing the Internet	-	5.1
Chatting	-	3.8
Video clips	-	4.3

The test persons were asked, using open questions, to say what their favourite MIES function was and what function(s) they missed. Favourite functions of both groups were access to the Internet and the ability of find locations with the navigation programme. The conference visitors also mentioned as a favourite the up to date conference information while the students liked the chat function. Functions mentioned that could be improved were the automatic functioning of the login settings and the MIES position updater, the integration of the Internet browser and the navigation programme, the possibility to update personal information for example to personalise the conference schedule, content richness for example on health care and museums. Some recommendations to improve the navigation were also provided.

The general impression is that the test persons were quite positive about MIES. Despite the fact that there were some technical problems, not all the features worked as well as they were intended to and the test persons had to carry a bag with the devices and chargers, the users loved playing with the equipment; the possibility to access the Internet anyplace, anytime especially was consistently mentioned as a good functionality. The TomTom navigator was found to be a favourite function; this was a surprise since the programme needs some time to get to know it.

7.6.2 Partners' experiences

The results of the interviews with 12 partners and participant-observation provided information to gain an understanding of the partner's point of view regarding the value network related issues. It was evident that there should be a win-win situation for all participants. In the long run this win has to be expressed in a financial way. In the short term value is gained from improving actors' image and gaining experience. One respondent remarked that there are partners with an established reputation, but that this might be a reputation with which you do not want to be associated. E.g. when a small company wants to establish an innovative, entrepreneurial image, it might not want to be associated with a (large) firm with an established but conservative image. All the firms involved were asked what the benefit was for them to participate. In the short term earning revenues was not an option for any of the partners since they were not paid and a market launch was not planned. Nevertheless they saw other benefits that made them decide to participate. The research opportunities were the major reason for the research organisations. Reasons such as getting experience, image and thus revenues in the long run were important for the corporate partners. An important reason to participate for the consultancy firm The Vision Web was that T-Mobile is one of their most important customers. Reasons to participate for the partners from the TU Delft were mainly research opportunities and chances to learn, gain experience, and also – to some extent – to build a reputation.

The involved partners strongly agreed that it is important to reduce uncertainty when creating a value network by using embedded relational ties. They argued that especially in innovative projects under time pressure it is important to know the right people if you are searching for some knowledge.

In Table 7-6 is described to what extent the planned design activities were executed satisfactorily, not executed satisfactorily or were not executed at all.

Table 7-6 Confrontation prescriptive conceptual and empirical model

Nr*	**Activity**	**Done**	**Remark**
A.O.1	Process Design	Yes	The actors' scope rested heavily on the embedded relational ties. Remarks for organising the project management: plan go/no go moments more explicitly keep track of design decisions made, make them explicit and communicate them to all involved actors defining roles, responsibilities and tasks is important in every phase.
A.O.2	Agreeing on intention and external communication	Yes	Besides agreeing on participation it is already necessary in this phase to make clear who is going to communicate what externally, to the press etc. The mobile operator especially required communication contracts.
A.T.1	Scan available technology	Yes	The available technology was to a great extent determined by the mobile operator.
A.T.2	Overview functional/technical design decisions	Not explicitly enough	The storylines were not specific enough as a basis for functional requirements. One of the important lessons learnt is that the functional requirements should have been clearer. One supplier mentioned that technical requirements are not formulated in projects under time pressure. He advocated that the only thing that counts is to make it work and people use the trial and error method.
A.S.1	User requirement elicitation	Yes	A very important activity. This activity was academically very interesting; we published a paper on this activity and got ideas for further work. A remark on activity A.S.1 was that users should get feedback on what is done with the outcome of the GDR sessions. Asking for feedback on design decisions for users would do this. This should be added as a process control activity.
A.S.2	Analysing existing services	Yes	Navigation designers investigated other visitor/tourists sites before they started their design.

Nr*	Activity	Done	Remark
A.S.3	Drawing story-boards	Yes	We made storylines with little stories and pictures. We had leaflets and posters with these storylines and this worked really well.
P.O.1	Shared requirement session	Yes/No	A launch meeting to get to know each other is important. We organised such a meeting with drinks. The planned collaborative service design session with partners was not executed. One of the partners mentioned that he would have liked a GDR session as done with the users but then with all the partners. This confirms that activity P.O.1 including a collaborative service design session should have taken place.
P.O.2	Business case for each partner	Yes/No	The partners did this more implicitly. All had their own reasons to participate. T-mobile had a planned budget for the whole UMTS test-bed. The investments of the other partners were relatively low and they decided to spend time or money when a decision had to be made.
P.O.3	Defining roles, responsibilities and tasks	Yes/No	The project initiator took care of the coordination of the activities to keep the network intact. This also included social activities like a launch meeting and a dinner at the end. The governance model to manage the content providers can be described as a 'thin-walled-garden' model managed by the Service Provider. The project leader for technology was explicitly appointed and he organised weekly sessions to discuss design decisions and the progress of the project. A service project leader was missing and critical comments on this were made during the interviews.
P.T.1	Technical architecture design	Discussion	Figure 7-3 was the object of discussion for some weeks and later people also mentioned that they had missed a more detailed overview.

Nr*	Activity	Done	Remark
P.T.2	Selection of technical components	Yes	Not all components as mentioned in P.T.2, were built in this project. The billing system was not part of the project. Security also did not get explicit attention. Both these system components have to be taken care of in a further design round.
P.T.3	Installing platform	Not at first	This was not done properly. During activity S.T.1 different teams did the programming for the privacy and the localisation application. It turned out that they had used different software languages and the impact of this was that different servers were needed. The solution in the first version of MIES (for the ESS test) was that templates had to be copied and if one team changed something in the template the user saw differences in the MIES menu. This was of course not what was wanted. In the second version of MIES all applications were reprogrammed to the same language.
P.T.4	Application design	Yes	This was difficult for some designers because they missed a more detailed technical architecture design, see P.T.1.
P.S.1	Content design	Yes	The various content providers designed their part of the 'MIES on the campus portal'.
P.S.2/P.S.3/P.S.4	Navigation, User Interface, Test	Yes	The different content ingredients were integrated in this activity.
S.O.1	Detailed planning of roles, responsibilities and tasks	No	All the tasks were executed in an informally managed way.
S.O.2	Calculating benefits and costs	No	Too much time pressure.

Nr*	Activity	Done	Remark
S.T.1	Programming and scripting	Yes	Different persons designed different applications, see P.T.3. There were two kinds of comment on this activity related to 'writing documentation'. Some programmers said that this should be done automatically. There are programmes for software programming that automatically produce the documentation. Other partners argued that this is never done during work on a project under time pressure but it is a task that should be done afterwards.
S.T.2	Assembling and testing different components	Yes	One day all the programmers sat together and assembled all the building blocks and solved any problems that appeared.
S.T.3	Pre-testing prototype	Not enough	An important test for the mobile operator was to put all the UMTS devices on at the same time. This went well. Many technical problems appeared during the test person's test that could have been avoided by better technical testing of the applications.
S.S.1	Deciding on service quality and customer communication	Yes	This was done without problems for the purpose of the test in which branding of the service and the communication with the users could take place on a very low level. This activity requires much more attention in a second round of the design approach.
S.S.2	Privacy and legal aspects	Yes	Privacy was handled as a separate project, see (Ali Eldin & Wagenaar, 2004). Legal aspects concerning liability were taken care of.
I.O.1	Execution of tasks	Yes	All the tasks were executed in an informally managed way.
I.S.1	Organising the launch	Yes	No remark, activity executed without problems.
I.S.2	Distribution	Yes	No remark, activity executed without problems.
I.T.1	Preparing the equipment	Yes	Activity executed under heavy time pressure by very inventive students who solved all the problems creatively.

Nr*	Activity	Done	Remark
T.O.2	Analysis of the business case	No	The partners reported individually on the benefits they got out of the project. The first plan was to organise a meeting with all partners to discuss the joint business case for the service and to discuss how cost and revenues could be divided into the exploitation phase. However, since no exploitation plans were in place such a meeting was not worth the time and investment.
T.T.1	Support of developers during test	Yes	No remark, activity executed without problems.
T.T.2	Monitor usage and providing standby support	Yes	No remark, activity executed without problems.
T.S.1	Gathering user's experience	Yes	No remark, activity executed without problems.

* The numbers relate to the design activities as mentioned in chapter 6.

7.7 Evaluation

The project was considered a success by the users of the service and the organisations involved in the project. However, critical remarks can be made on the design process and on the further course of the project. In chapter 6 it was mentioned that the challenge is to find design methods and tools that provide answers to the trade-offs between the elements value network, technology and service concept. The challenges are discussed along the lines of these trade-offs.

The first trade-off concerns the service concept and the technology: 'How to get the enabling technology to support a service that fulfils the demand of the users?' The design method followed in this case has as starting point that the design process should start with the investigation of the wants and needs of the user. This is followed by an inventory of the available building blocks to assemble the service. In the project was found that the gap between the storylines and the programming required for '*MIES on the campus*' was too big. The overview of the technical functions and the technical architecture design was not clear enough for the programmers. The technical project leader argued that functional design and technical design can only be done later in circumstances of time pressure since you cannot build a service fast if you first have to build a detailed technical architecture. The discussion reflects the contrast between proponents and opponents of rapid application design. It is a challenge to find a way to do this satisfactorily during the design phase.

The second trade-off concerns the technology and value network 'How to get a value network in place to provide the necessary technology?' There is interdependency between the technology and the needed resources. In cases like this the resources can only be found in different companies. These companies might work with different standards. It was found that one company works with PHP platforms and another one with ASP platforms. Since the process of standardisation is not finished, and probably will never be, it is important to find a way to deal with two conflicting requirements. It is necessary to have an open platform to facilitate many actors who can add value adding services, but it is also necessary that their applications work together. The provided solution for this was a platform in which it is easy to add building blocks. However, in the project it became clear that the installation of the web services platform cannot be completely open and needs orchestration. This taught that it is very important to make a centralised decision on runtime environments.

The third trade-off concerns the value network and the service concept 'How to create a network of actors and coordinate the activities of these different actors to deliver value to the user?' In the project a combination of project management and process management principles were applied. An important issue in this trade-off is that all partners share the same perception on the customer value of the service. It was intended to hold a collaborative service design session with all partners, based on the EasyWinWin approach from Briggs and Gruenbacher (2002), at the start of the project. However, somehow it did not happen. This may have been due to bad time management ('there was no time') or due to a lack of interest on behalf of the partners. It would be worthwhile to do a follow up project with this activity explicitly included.

This project ended with the results of the prototype test in a pilot. The service was not further elaborated and launched on the market. Therefore, it was not possible to assess the business case. The test persons were asked how much they were prepared to pay for the service. This has to be assessed by estimating the costs necessary to develop and maintain service, the expected amount of users and what they are prepared to pay for the service. These revenues and costs are input for the revenue model, which shows how these costs and profits are divided among the participants. Such a revenue model must be negotiated with all the involved partners. This activity falls in the next round of the design approach.

The next round of the design approach should be initiated by a corporate partner and not the university. The mobile operator appeared not to be this actor since they argued that they offer the network and expect other service providers to develop services to run on the network. The local content provider in this case was not large enough to be this service provider. So there seems to be a gap in the market for an entrepreneur who wants to be in the service provider business. This means the whole round of the design approach needs to be repeated with emphasis around the revenue model of the 'MIES on the campus' service. Under what conditions do the partners want to launch this service on the market? Do the partners forecast a good business case for their role in the value network that offers the service? After the second round it should become clear if it is necessary to go through the design approach for a third time before the market launch.

Appendix A Unified Modelling Language diagrams

Modelling languages provide a vocabulary for stakeholders to communicate with each other. The Unified Modelling Language (UML) is designed to let system analyst, developers, programmers and maintainers view a system from different perspectives and in varying degrees of abstraction (Fowler. 2000). UML is a modelling language whose vocabulary can be used to both conceptually describe a service systems as well as specifying the methods and attributes of a software system at the programming code level. UML is often viewed as a de facto standard for modelling software systems and today computer science students are expected to know this language. UML is supported by many software programmes provided by different software vendors (e.g. MS Visio) and programmes even exist that are able to generate software code from UML diagrams (e.g. Rational Rose).

Models are of essential importance in the design process. They help to visualise concepts and specify them in more details. They can be used to communicate and specify design ideas. Moreover, they can be used for documenting the designed blueprint. An essential requirement for the use of models is that stakeholders are able to understand the modelling language. Communication of ideas often enables discussion about the design constraints and requirements and can help to generate creative ideas and in this way improve the design of the service system. Although UML is developed for designing software systems it can be helpful in designing service systems, especially in modelling the behaviour of users interacting with the service system.

UML consists of a collection of modelling notations that can be used to describe different views on the service system. One of those modelling notations, Use Case modelling, is of especial importance for the engineering of service systems as it specifies how users interacts with service systems and will be discussed next. In the situation that the service system is a software system other UML modelling notations can be used. Before we present Use Case diagrams for modelling the behaviour of service systems we provide an overview of UML diagrams commonly used. Theses include (e.g. Fowler. 2000):

Class Diagram. Class diagrams are widely used to describe the types of objects in a system and their relationships, using design elements such as classes, packages and objects. It also displays relationships between object classes such as containment, inheritance, and associations.

Use Case Diagram. The Use Case diagram displays the relationship among actors (types of users or roles) and "Use Cases". Use Cases are typical situations in which the system that is designed, can be used.

Interaction Diagram .Two different types of interaction diagrams exist within UML:

1. Sequence Diagram displays the time sequence of the objects participating in the interaction. This consists of the vertical dimension (time) and horizontal dimension (different objects).
2. Collaboration Diagram displays an interaction organised around the objects and their links to one another. Numbers are used to show the sequence of messages.

State Diagram. The State Diagram displays the sequences of states that an object of an interaction goes through during its life in response to received stimuli, together with its responses and actions. It provides a detailed description of the life cycle of each object class.

Activity Diagram. The Activity Diagram displays a special state diagram where most of the states are action states and most of the transitions are triggered by completion of the actions in the source states. This diagram focuses on flows driven by internal processing.

Physical Diagrams. The physical diagrams give more detail about the implementation of the system, and two different types can be distinguished:

3. Component Diagram displays the high level packaged structure of the code itself. Dependencies among components are shown, including source code components, binary code components, and executable components. Some components exist at compile time, at link time, at run times as well as at more than one time.
4. Deployment Diagram displays the configuration of run-time processing elements and the software components, processes, and objects that live on them. Software component instances represent run-time manifestations of code units.

In the following section we discuss the use of one UML modelling notation for modelling service systems.

The Use Case modelling notation

The mobile service systems are operated by all kind of actors from various organisations. Actors can include the end-user of a service, users updating the content of a system, administrators and so on. The Use Case modelling notation described how actors interact with the elements of a service system. A *Use Case* is a description of a set of actions that a system performs and results in some observable result for the actors. In this way Use Cases structure the behavioural of a system from the user point of view. As such, Use Cases specifies the behaviour of a (part of) a systems, and a description of a typical set of sequence of actions that a system performs.

Use Case modelling helps with reasoning about how the service system will be used. A service system can be used in many ways and consequently many sequences of activities are possible. A particular sequence of behaviour describing how a user interacts with the service systems is called a *Use Case scenario*, or scenario for short. Scenarios can be developed to detail the different paths of events and activities for each Use Case. Actors can often use a service system in almost an infinite number of ways. Therefore Use Case descriptions are often limited to a number of typical scenarios, called primary scenarios. The *primary scenario* describes the typical case where nothing goes wrong. In a typical design of a service system many scenarios are included.

Use Case modelling is a technique often used by developers in interactions with the possible users of the service system. As such it is a communication vehicle between users and developers. By making use of modelling technique, system developers can capture the intended behaviour of the service system they are developing without having to specify how the behaviour is implemented. During the initial stage of a project the typical scenarios should be defined. As the project continues non-typical scenarios and exceptions might be needed to be modelled and more Use Case models might become necessary. Moreover, as the design process continues often Use Cases are developed that are specialised version of the initial Use Case scenarios. Also Use Case diagrams might not only be applied to model the complete system, but also to parts of the system.

Use Case models are sets of scenarios describing the interaction between a user and a system. The Use Case Model describes the proposed functionality of the service system from an actor point of view. *A Use Case model represents a discrete unit of interaction between a user (human or machine) and the system* (Fowler, 2000). The two main components of a Use Case diagram are Use Cases and actors as shown in Figure A-1. The Use Case specifies how the service system should behave without having to specify how the Use Case will be implemented. For example someone would like to describe the typical scenario to transfer money using a banking system. The actor is the customer of the banking service system who wants to transfer money from one account to another account. Examples of Use Cases are login transfer money, select amount and account, confirm transfer and log out. The Use Case model describes the sequences, in which each sequence represents the interaction of the user outside the system (the person who wants to transfer the money) with the service systems.

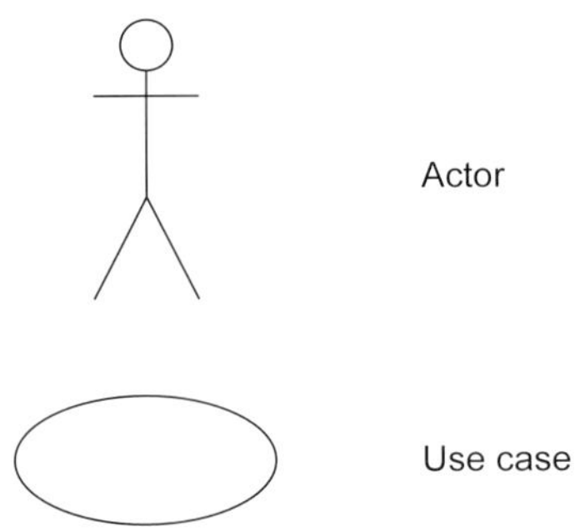

Figure A - 1. Use Case notations

Within the Use Case notation a *Use Case* is a functionality performed by the service system, and presents an external view of the system that represents some action the user might perform in order to complete a task. A Use Case is a single unit of meaningful work. Each Use Case contains some functionality that should be included in the proposed service system. Every Use Case must be given a unique name in order to distinguish it from other Use Cases. As a guideline Use Cases are *short active verb phrases* naming some behaviour. The name is often derived from the vocabulary of the service system that needs to be designed, i.e. the typical naming by persons is used. This can avoid conflicts and misunderstanding.

Actors can carry out Use Cases. An *actor* represents a coherent set of roles that users of Use Cases play. In our example this could be "customer". An actor is not necessarily a human being; it can be another system that will interact with the system you are modelling. Actors are not part of the systems and should be positioned outside the system. Actors describe the roles that human *or* other information systems can play. A user is often described by several roles. For example a customer can have a transfer money role, deposit money role, and manage account role.

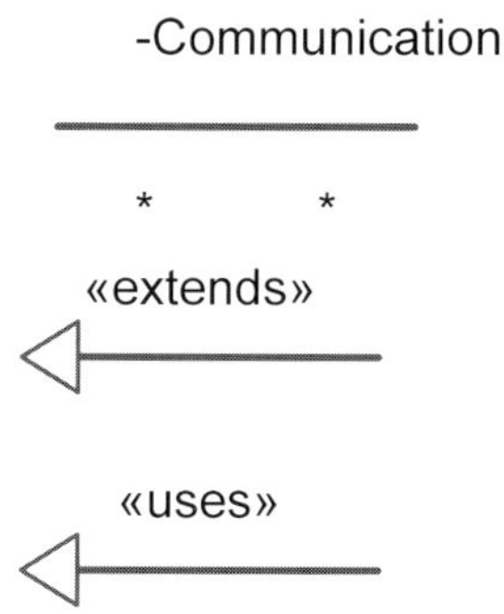

Figure A - 2. Use Case interaction notations

There are three types of relationships in a Use Case diagram as shown in Figure A - 2.. The <<*communication*>> relationship visualises the relationship between Actors and Use Cases. The <<*extend*>> relationship is used when one Use Case is similar to another Use Case but does have more functionality, for example grade sufficient degree extends grading. The <<*uses*>> relationship occurs when functionality is similar across more than one Use Case and you do not want to keep copying the description over again. In this way reuse of functionality is enabled, for example buying uses credit card check.

The behaviour of a Use Case scenario is described by a flow of events. It should be clearly defined how the Use Case scenario is initiated and when it starts and ends. The list of actions starts at the top of a Use Case model and ends at the bottom of the model. Often Use Cases are derived from interviews or from written text like reports, memos and so on. A number of steps can be taken to develop a Use Case model.

1. Identify the actors of the service system. These are often denoted by the subject in a sentence.
2. Organise the actors by identifying general (e.g. customer role) and specialised roles (transfer money role). Often the specialised roles result in a typical scenario.
3. Identify the typical Use Case scenarios. This can be used by looking the primary ways in which actors can operate the service system and also by defining a clear goal that should be accomplished.
4. Determine the start and the end of the typical Use Case scenario. If a Use Case scenario grows too complex, it can be split up into subsequent scenarios.

5. Identify the Use Cases in each of the scenarios. Describe the flow clearly enough, so stakeholders involved and non-involved are able to understand it. Focus on what the Use Case does and not on how it does it.
6. Determine the systems functionality that might be reused and model this using the <<*uses*>> relationship.
7. Determine exceptional ways and model this using <<*extend*>> relationship.
8. Draw the Use Case model. In general show only those Use Cases that are important to understand the complex behaviour of the service system. Rare or exceptional Use Cases might be omitted, however, they should be described in the additional list of requirements. The Use Case model should have clear defined system boundaries and only show actors related to the Use Case.

The steps are often not performed sequentially and may contain many loops and iterations. Often a Use Case scenario might become more complex over time as more and more behaviour is added. Then the Use Case scenario can be split up into multiple scenarios. It is unusual that only one Use Case scenario needs to be made.

A Use Case Modelling Example

We will demonstrate these steps using the text below derived from several interviews with potential users of the system. When reading the text the Use Case developer can start with marking the words that denote to actors (often the subject) and to Use Cases (often verbs).

A *customer* wants to buy a service and therefore needs to transfer money to another account. The customer starts its computer and a web browser. If the user has no account it might contact the account manager of the bank to create a new account. The Use Case starts when the systems prompts for a login name and password. The customer *logs in* and the system shows the menu. The customer *selects* the transfer money options and a dynamic form appears. The customer should *fill in* the amount and the account number of the account the money should be transferred to. After sending the form over a secure connection, the information the service system *checks* if there is sufficient money in the account available. A confirmation screen appears and the customer should either *confirm* or *cancel* the transaction.

When modelling the Use Case it is of crucial importance to focus on what the Use Case does, not on how it does it. The main actor in this scenario is the customer who wants to transfer money. The modelling of Use Cases starts by listing a sequence of steps a user might take in order to complete an action. The beginning and end are already clear from the text. The Use Cases can be identified by looking at the verbs and marking them. For example a user transferring money might follow these steps as shown in the following figure.

Use Case scenario name:	Transfer money
Use Case scenario intent:	To enable a customer to transfer money from one account to another account to pay for a service
Version number	1.0
Use Case scenario description	
1. Log in using a username and password 2. Select money transfer option from the menu 3. Fill in amount of money that needs to be transfer and the account number of the destination 4. Either Confirm or cancel money transfer 4a Confirm the money transfer 4b Cancel the money transfer 5. Log out	
Requirements and other information	
The whole session should be secured The account check function should be reused	

Figure A - 3. Description of the Money transfer scenario for the banking service system

Often other roles are involved, for example, an account manager who is able to create a new account for a non-existing customer. Also the account might be destroyed. This results in the money transfer scenario for the banking service systems as shown below. It might become clear that the graphical representation is much easier to understand and to communicate that the written Use Case description. In the case studies description in this book many more examples of Use Case descriptions can be found.

Figure A - 4. Money transfer scenario for the banking service system

The banking service systems can be completely described by modelling several typical scenarios. These scenarios might also re-use the account check functionality, as this is described using a <<uses>> relationship. The Use Case models can be used for the following purposes:

1. Model a service system to a degree sufficient for developers to programme the software, without going into too many details which would make it more difficult to understand.
2. Communicating the behaviour to potential users. Use Cases are relatively easy-to-understand. By specifying classes, methods and attributes users might be lost in details instead of understanding the main system behaviour. The communication process might even result in suggestions fro improvement of the Use Case model.
3. Understand the complex relationship between the elements of a system. For example it is relatively easy to determine the building blocks that should be made generic to enable reuse in multiple situations.

4. Supporting testing. Typical user scenarios might be followed to test the systems if it satisfies requirements.
5. Ensuring adaptability. Not only the current behaviour might be described, but also scenarios might be developed describing potential future behaviour of the system that might be added later. In this way the system might become more sustainable and have a longer life-cycle.

The Use Case description can be completed by specifying the non-functional requirements, constraints and other information that might be used when implementing the service system. Constraints are the formal rules and limitations that a Use Case operates under, and includes pre- post- and invariant conditions. A pre-condition specifies what must have already occurred or be in place before the Use Case may start. A post-condition specifies what will be true once the Use Case is complete. An invariant specifies what will be true throughout the time the Use Case operates. Also information describing exceptional situations or expected problems when implementing the service system might be included.

Design example: Cycling tours service

The cycling tour service provides personalised cycling routes to mobile users. Based on the preferences of the user, including cultural interests, preferred types of cycling routes, and favourite cycling heroes, the service provides a cycling route. During the route, the user will receive cultural information, information about the performance of his or her cycling heroes, nearby hotels and restaurants, and instructions on the route itself. This design example emphasises the technical architecture of the service system, and illustrates the use of use cases in service system design.

To access the service, the user will need a mobile phone and a GPS. Before using the service for the first time, the user has to fill in some personal information. When the user requests a cycling tour, the service determines where the user is, using the GPS. Then, based on the preferred endpoint and the predefined information about the user, the service determines an interesting cycling route. This route has to be confirmed by the user, and is then presented to him/her on the screen of the mobile device. During the tour, the user will also receive spoken directions.

During the tour, the service will use the positioning capabilities of the GPS chip to determine the speed of the user. The service compares this to historical data about the user's cycling heroes and shows this to the user. Based on the interests of the user, information about cultural sites is also given during the tour by voice messages. When such a message comes in, the cyclist can choose to stop and request more information about the cultural point of interest from the service. During the tour, the user can also request information about nearby hotels and restaurants, which will be returned on the screen of the device. The interactions of the user with the service system are visualised in a use case diagram, see **Figure A -5**.

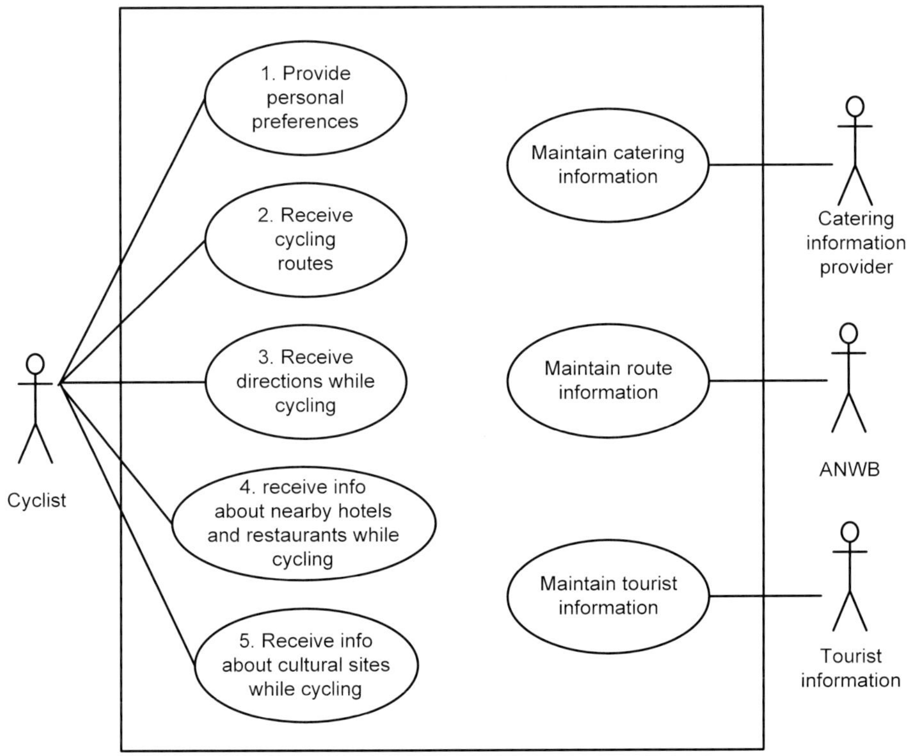

Figure A -5. Use case diagram for 'Cycling tours service' design example

We will now describe each use case in more detail.

1. *Provide personal information:* the user enters his/her personal preferences via the mobile device. The mobile device sends these preferences to the service system, and there the information is stored in a database.
2. *Receive cycling routes:* the user requests a cycling route via the mobile device. Optionally, the user can choose to enter the preferred duration of the tour and the preferred end location. The mobile device then uses the GPS chip to determine the location of the user. The coordinates, the duration and end location are now sent to the service system.

The service system uses the information that has been sent by the device and the information about personal preferences of the user that is stored in the database to calculate possible routes. It also uses information from the Algemene Nederlandse Wielrijders bond (ANWB, the Dutch Touring Club) in this step. This organisation provides information about cycling routes in Europe. The service system will also use the services of a GIS provider in this step, in order to calculate the routes. Once the service system has calculated the routes, they are sent to the mobile device. The user can browse through the routes and select which one he/she would like to take.

3. *Receive directions while cycling:* while cycling, the mobile device determines the position of the user with the GPS chip. This positioning information is periodically sent to the service system, which uses it to send instructions on how to drive to the user. The service system will use the services of the GIS provider to generate the driving instructions.
4. *Receive information about nearby hotels and restaurants while cycling:* while cycling the user requests information about nearby hotels and restaurants. The user requests this via the mobile phone, and the mobile phone sends the current coordinates to the service system. Now, the service system uses the coordinates, the information about the preferences of the mobile user and information about hotels and restaurants to give an advice. This information about hotels and restaurants will be provided by a catering information provider. The service system will again use the services of the GIS provider to compare the location of the user to the location of the hotels and restaurants. Information about the selected hotels or restaurants is sent back to the mobile device, and the user can read the information on the screen.
5. *Receive information about cultural sites while cycling:* as in the third use case, the coordinates of the user's position are periodically sent to the service system. The service system uses the services of the GIS provider to compare these coordinates to the position of cultural points of interest. When the service system sees that the user is passing a cultural point of interest, it looks up the preferences of the user to determine whether the user will like it. If so, it sends spoken information about the cultural heritage to the mobile device. The mobile device now reads or displays the announcement.

If the user becomes curious as a result of the announcement, he/she can stop cycling and request more information about the cultural site via the mobile phone. This request is then sent to the service system, which returns additional information. Information about cultural heritage will be retrieved from the ANWB.

Index

Glossary

3G Third Generation

ASP Applications Service Provider

ASP Active Server Pages

CBD Component Based Development

CDMA Code Division Multiple Access is a technology for digital transmission of radio signals between a mobile telephone and a radio base station

DPEC Directive on Privacy and Electronic Communication of the European Union

DTD Document Type Definition, a language for defining a document structure

E-service an activity or series of activities of intangible nature that take place in interaction through an Internet channel between customers and service employees or systems of the service provider, which are provided as solutions to customer problems, add value and create customer satisfaction (based on the definitions of Grönroos (2001) and (Kasper et al., 1999))

GIS Geographical Information Systems

GPRS General Packet Radio Service is a standard for cellular phones

GPS Global Positioning System can determine the geographical position

GSM Global System for Mobile communication is a standard for cellular phones.

GSS Group Support Systems

HTML HyperText Markup Language is a mark-up language with a fixed set of tags that allows users to specify the appearance of a document.

I-centric individual centric

ICT Information and Communication Technology

Infomediary information intermediary

ISP Internet Service Providers

IP Internet Protocol

IT Information Technology

LAN Local Area Networks are networks covering a certain location such as a business.

LBMS location-based mobile services

LBS location-based services

LDAP Lightweight Directory Access Protocol

m-value mobile value

MAN Metropolitan Area Networks are networks covering a metropolitan wide area.

Middleware Technology enabling the exchange of information between information systems and encapsulating implementation details

MIES Mobile Information and Entertainment Services

Mobile information service an activity or series of activities of intangible nature that occur when the consumer is mobile and a mobile telecommunications network supports the interaction through an Internet channel between customers and service employees or systems of a service provider. The aim is to provide solutions to customer problems and needs, add value, and thus create customer satisfaction

Mobile service system a group of components that work together for delivering (series of) activities of intangible nature and with added value for a mobile customer using a mobile network

MPU Mobile Position Updater

NSP Network Service Provider are often major telecommunication companies that maintain and service the Internet backbone.

Operational processes the activities and their interdependencies that directly contribute to the development and delivery of a service

Organisational network provides the co-ordination between the resources, such as people and knowledge, in a (set of) service organisation(s)

PDA Personal Digital Assistant

PHP Hypertext Pre-processor (recursive acronym)

PSTN Public Switched Telephone Network

QoS Quality of Service

RBV Resource-based view

Service concept the strategy with which a service proposes to create value for its customers

SOA Service Oriented Architecture is an architectural style, according to which application functionality is not provided by one large monolithic application, but is provided by services that can be combined to get the required functionality.

SOAP Simple Object Access Protocol) provides an envelope around XML message in order to exchange structured information

SWS Semantic web services refer to the joint use of web service technology and semantic technology

TCE transaction cost economics

Technical architecture the description and prescription of a system consisting of components, their relationship to each other and the environment accompanied with standards and principles guiding the design and development.

TPM Technology, Policy and Management (Faculty of)

UDDI Universal Description, Discovery and Integration is A 'yellow pages' of services

UML Unified Modelling Language, designed to let developers and customers view a software system from a different perspective and in varying degrees of abstraction

UMTS Universal Mobile Telecommunications System is a standard for cellular phones

Use Case a representation of a discrete unit of interaction between a user (human or machine) and the system (Fowler & Scott, 2000)

VPN Virtual private network that uses the public Internet to carry information but remain private by using encryption, access and security protocols.

WAP Wireless Application Protocol

Web Services Web services is a technology that enables the provisioning of functionality, on an application level or on a business level, by means of a standardised interface in a way that they are easily invoked via Internet-protocols.

Web Service choreography Web service choreography defines the sequence and conditions of public message exchanges between multiple web services (www.w3c.org).

Web service orchestration Web service orchestration is the process of invoking internal and external web services from a predefined process flow that is executed by an orchestration engine (www.w3c.org).

Wifi Wireless fidelity is wireless standard for local area networks

WiMax Worldwide Interoperability for Microwave Access is a wireless standard for making broadband connections

WS see Web Services

XML Extensible Mark-up Language, a language for describing documents

XSL Extensible Style Sheet Language is a language for formatting documents

References

Afuah, A., & Tucci, C. L. (2001). *Internet business models and strategies: text and cases.* Boston: McGraw-Hill/Irwin.

Alam, I., & Perry, C. (2002). A customer-oriented new service development process. *Journal of Services Marketing, 16*(2), 515-534.

Ali Eldin, A., & Wagenaar, R. W. (2004, June 14-16). *A Fuzzy Logic based Approach to support users self-control of their private contextual data retrieval.* Paper presented at the 12th European Conference on Information Systems (ECIS04), Turku, Finland.

Allen, P. (2001). *Realizing e-Business with Components.* Harlow, UK: Pearson Education.

Allen, P., & Frost, S. (1998). *Component-Based Development for Enterprise Systems.* Cambridge, UK: Cambridge University Press.

Alter, S. (2002). *Information Systems* (fourth ed.). New Jersey: Pearson Education.

Amaoko-Gyampah, K., & Salam, A. F. (2004). An extension of the technology acceptance model in an ERP implementation environment. *Information and Management, 41*(6), 731 - 745.

Amit, R., & Zott, C. (2001). Value creation in e-business. *Strategic Management Journal, 22*(6-7), 493-520.

Anckar, B., & D'Incau, D. (2002). *Value-Added Services in Mobile Commerce: An Analytical Framework and Empirical Findings from a National Consumer Survey.* Paper presented at the Proceedings of the 35th Hawaii International conference on System Sciences, Big Island, Hawaii.

Andriessen, J. H. E. (2003). *Working with groupware: understanding and evaluating collaboration technology.* London: Springer.

Appelman, J. H., & Van Driel, J. (2005, 3-6 January). *Crisis-response in the Port of Rotterdam: can we do without a facilitator in distributed settings?* Paper presented at the Proceedings of the 38th Hawaii International Conference on System Sciences, Big Island, Hawaii.

Arsanjani, A. (2002). Developing and Integrating enterprise components and services. *Communications of the ACM, 45*(10), 31-34.

Axelsson, B. (1987). Supplier Management and Technological Development. In H. Håkansson (Ed.), *Industrial and technological development; a network approach* (pp. 128-176). London: Croom Helm.

Baldi, S., & Pyu-Pyu Thaung, H. (2002). The Entertaining Way to M-commerce: Japan's Approach to the Mobile Internet - A Model for Europe? *Electronic Markets, 12*(1), 6-13.

Ballon, P., & Hawkins, R. (2003). *From Business Models to Value Networks* (Working Paper). Delft, The Netherlands: TNO/STB.

Banks, J. (1998). *Handbook of Simulation: Proncíples, methodology, advances, applications, and practice.* New York: Wiley.

Bannon, L. J. (1991). From human factors to human actors: The role of psychology and human-computer interaction studies in system design. In J. Greenbaum & M. Kyng (Eds.), *Design at work: Cooperative design of computer systems* (pp. 25-44). Hillsdale, NJ: Lawrence Erlbaum.

Barnes, S. J. (2003). *Mbusiness. The strategic implications of wireless communications.* Oxford: Elsevier Butterworth-Heinemann.

Bateson, J. E. G. (1992). *Managing Services Marketing.* Orlando, Fl: The Dryden Press.

Benbasat, I., & Lim, L. H. (1993). The effects of group, task, context, and technology variables on the usefulness of group support systems: A meta-analysis of experimental studies. *Small Group Research, 24*(4), 430-462.

Bi, G., Zysman, G. I., & Menkes, H. (2001). Wireless Mobile Communications at the Start of the 21st Century. *IEEE Communications Magazine, 39*(1), 110-116.

Binnekamp, R., van Gunsteren, L. A., & van Loon, P. (2006). *Open Design, a Stakeholder-oriented Approach in Architecture, Urban Planning, and Project Management.* Amsterdam: IOS Press.

Blanchard, B. S., & Fabrycky, W. J. (2006). *Systems Engineering and Analysis* (4th ed.). Upper Saddle River, NJ: Pearson Prentice Hall.

Blom, J., Chipchase, J., & Lehikoinen, J. (2005). Contextual and Cultural Challenges. *Communications of the ACM, 48*(7), 37-41.

Boehm, B. W. (1988). A Spiral Model of Software Development and Enhancement. *IEEE Computer, 21*(5), 61-72.

Boer, C. A. (2005). *Distributed Simulation in Industry.* Unpublished Ph.D., Erasmus University, Rotterdam.

Boer, C. A., & Verbraeck, A. (2003). *Distributed Simulation with COTS Simulation Packages.* Paper presented at the Proceedings of the 2003 Winter Simulation Conference.

Booch, G. (1991). *Object Oriented Design with Service systems.* Redwood City CA: Benjamin / Cummings.

Briggs, R. O., De Vreede, G.-J., & Nunamaker, J. F., Jr. (2003). Collaboration Engineering with ThinkLets to Pursue Sustained Success with Group Support Systems. *Journal of Management Information Systems, 19*(4), 31-64.

Briggs, R. O., & Grünbacher, P. (2002). *EasyWinWin: Managing Complexity in Requirements Negotiation with GSS*. Paper presented at the Proceedings of the 35th Hawaii International Conference on System Sciences, Big Island, Hawaii.

Briggs, R. O., Nunamaker, J. F., Jr., & Tobey, D. (2001). *The Technology Transition Model: A Key to Self-Sustaining and Growing Communities of GSS Users*. Paper presented at the Proceedings of the 34th Hawaii International Conference on System Sciences, Big Island, Hawaii.

Brown, A. W. (2000). *Large-scale component-based development*. Upper Saddle River, NJ: Prentice Hall.

Brown, J. E., & Hendry, C. (1998). Industrial Districts and Supply Chains as Vehicles for Managerial and Organizational Learning. *International Studies of Management and Organization, 27*(4), 127-157.

Bruseberg, A., & McDonagh, D. (2003). Organising and Conducting a Focus Group: The Logistics. In J. Langford & D. McDonagh (Eds.), *Focus Groups, Supporting Effective Product Development* (pp. 21-45). London: Taylor & Francis.

Buede, D. M. (2000). *The Engineering Design of Systems: Models and Methods*. New York: Wiley.

Buijs, J., & Valkenburg, R. (1996). *Integrale produktontwikkeling*. Utrecht, the Netherlands: Lemma.

Cadle, J., & Yeates, D. (2001). *Project management for information systems* (Third ed.). Harlow: Financial Times/Prentice Hall.

Camponovo, G., & Cerutti, D. (2004, 12-13 July). *The SPAM issue in mobile business; a comparative regulatory overview*. Paper presented at the Proceedings of the Third International Conference on Mobile Business, New York.

Carlsson, C. (2003). *Smart Adaptive Mobile Life Enhancements; Proposal FP6-2003-IST-2*. Abo: Abo Akademi University.

Checkland, P. (1993). *Systems Thinking, Systems Practice*. Chichester: Wiley.

Checkland, P. (1999). *Soft Systems Methodology: a 30-year retroperspective*. Chichester: John Wiley & Sons Inc.

Checkland, P., & Holwell, S. (1998). *Information, Systems and Information Systems. Making sense of the field*. Chichester: Wiley.

Checkland, P., & Scholes, J. (1990). *Soft Systems Methodology in Action*. Chichester: Wiley.

Christensen, C. M., & Rosenbloom, R. S. (1995). Explaining the attacker's advantage: Technological paradigms, organizational dynamics, and the value network. *Research Policy, 24*(2), 233-258.

Clarke, I. (2001). Emerging Value Propositions for M-commerce. *Journal of Business Strategies, 18*(2), 133-148.

Cohen, L. (1995). *Quality Function Deployment; How to make QFD work for you.* Reading: Addison-Wesley.

Crabtree, A. (2003). *Designing collaborative systems: A practical guide to ethnography.* . London, UK: Springer.

Creveling, C. M., Slutsky, J. L., & Antis, D. j. (2003). *Design for Six Sigma in technology and product development* (768 ed.). New Yersey: Prentice Hall.

Das, T. K., & Teng, B. S. (1998). Resource and Risk Management in the Strategic Alliance Making Process. *Journal of Management, 24*(1), 21-42.

Das, T. K., & Teng, B. S. (2000). A Resource-based Theory of Strategic Alliances. *Journal of Management, 26*(1), 31-61.

Davis, F. D. (1989). Perceived Usefulness, Perceived Ease of Use and User Acceptance of Information Technology. *MIS Quarterly, 13*(3), 319-340.

Davis, F. D. (1993). User acceptance of information technology: system characteristics, user perceptions and behavioral impacts. *International Journal of Man-Machine Studies, 38*(3), 475-487.

De Bruijn, H., Ten Heuvelhof, E., & In 't Veld, R. (2002). *Process management: why project management fails in complex decision making processes.* Dordrecht, the Netherlands: Kluwer Academic Publishers.

Deighton, N. (2003). *Managing and Measuring Mobile Workers.* Paper presented at the Gartner Symposium ITXPO.

Den Hengst, M., & De Vreede, G.-J. (2004). Collaborative Business Engineering: A Decade of Lessons from the Field. *Journal of Management Information System, 20*(4), 85-113.

Den Hengst, M., Van de Kar, E. A. M., & Appelman, J. H. (2004, 5-8 January). *Designing Mobile Information Services: User Requirements Elicitation with GSS Design and Application of a Repeatable Process.* Paper presented at the Proceedings of the 37th Annual Hawaii International Conference on System Sciences, Big Island, Hawaii.

Dennis, A. R., Valacich, J. S., & Nunamaker, J. F., Jr. (1990). An experimental investigation of the effects of group size in an electronic meeting environment. *IEEE Transactions on Systems, Man and Cybernetics, 20*(5), 1049-1057.

Ebers, M. (1999). The Dynamics of Inter-Organizational Relationships. In S. B. Andrews & D. Knoke (Eds.), *Network in and around Organizations* (Vol. 16, pp. 31-56). Stamford, Connecticut: Jai Press Inc.

ETSI. (2001). ETSI TS 101 723 v7.3.0. Digital cellular telecommunications system (Phase 2+); Locatoin Services (LCS); Service description, Stage 1. 3GPP TS 02.71 version 7.3.9 Release 1998 (2001-03). ETSI 2001. Retrieved 12 October, 2004, from http://www.etsi.org

Faber, E., Ballon, P., Bouwman, H., Haaker, T., Rietkerk, O., & Steen, M. (2003). *Designing business models for mobile ICT services.* Paper presented at the 16th Bled Electronic Commerce Conference eTransformation, Bled, Slovenia.

Fingar, P., Read, D., & Stikeleather, J. (1996). *Next Generation Computing: Distributed objects for business.* New York, NY: SIGS.

Fokkema, J. T. (2007). Preface. Delft University of Technology.

Forrester, J. W. (1968). *Principles of Systems.* Portland, OR: Productivity Press.

Fowler, M., & Scott, K. (2000). *UML distilled: a brief guide to the standard object modeling language.* Reading, Massachusetts: Addison-Wesley.

Fremantle, P., Weerawarana, S., & Khalaf, R. (2002). Enterprise services. *Communications of the ACM, 45*(10), 77-82.

Gadde, L. E., Huemer, L., & Håkansson, H. (2003). Strategizing in industrial networks. *Industrial Marketing Management, 32*, 357-364.

Gebauer, J., & Shaw, M. J. (2004). Success Factors and Impacts of Mobile Business Applications: Results from a Mobile e-Procurement Study. *International Journal of Electronic Commerce, 8*(3), 19-41.

Giaglis, G. M., Kourouthanassis, P., & Tsamakos, A. (2003). Towards a Classification Framework for Mobile Location Services. In B. E. Mennecke & T. J. Strader (Eds.), *Mobile commerce: technology, theory and applications* (pp. 67-85). Hershey: Idea Group Inc.

Gorlenko, L., & Merrick, R. (2003). No wires: attached: Usability challenges in the connected mobile world. *IBM Systems Journal, 42*(4), 639-651.

Granovetter, M. (1985). Economic Action and Social Structure: The Problem of Embeddedness. *American Journal of Sociology, 91*(3), 481-510.

Grönroos, C. (2001). *Service Management and Marketing. A customer relationship management approach* (2nd ed.). Chichester: Wiley.

Haanes, K., & Fjeldstad, O. (2000). Linking intangible resources and competition. *European Management Journal, 18*(1), 55-62.

Haartsen, J. C. (2000). The Bluetooth Radio System. *IEEE Personal Communications, 7*(1), 28-36.

Hagel III, J., & Singer, M. (1999). *Net worth, shaping markets when customers make the rules.* Boston: Harvard Business School Press.

Håkansson, H. (1987). *Industrial and technological development; a network approach.* London: Croom Helm.

Hedman, J., & Kalling, T. (2003). The business model concept: theoretical underpinnings and empirical illustrations. *European Journal of Information Systems, 12*, 49-59.

Heineman, G. T., & Councill, W. T. (2001). *Component-Based Software Engineering – Putting the pieces together.* Boston MA: Addison-Wesley.

Hertog, P. d., Broersma, L., & Ark, B. v. (2003). On the soft side of innovation: service innovation and its policy implication. *De Economist, 151*, 433-452.

Herzum, P., & Sims, O. (2000). *Business Component Factory – A comprehensive overview of component-based development for the enterprise.* New York, NY: Wiley / OMG Press.

Hite, J. M., & Hesterly, W. S. (2001). The evolution of firm networks: from emergence to early growth of the firm. *Strategic Management Journal, 22*(3), 275-286.

Holtzblatt, K., & Beyer, K. (1996). Contextual design: Principles and practice. In D. Wixon & J. Ramey (Eds.), *Field methods casebook for software design* (pp. 301-333). New York: John Wiley & Sons.

Ince, D. (2004). *Developing distributed and E-commerce applications* (2nd edition ed.). Harlow, Essex: Addison-Wesley.

Isaacs, E., & Walendowski, A. (2002). *Designing from both sides of the screen: how designers and engineers can collaborate to build cooperative technology.* Indianapolis: New Riders.

Itami, H., & Roehl, T. W. (1987). *Mobilizing Invisible Assets.* Cambridge, Mass: Harvard University Press.

Jackson, M. C. (1991). *Systems Methodology for the Management Sciences.* New York: Plenum Press.

Jacobson, I., Ericsson, M., & Jacobson, A. (1995). *The Object Advantage - Business process reengineering with object technology.* New York: ACM Press / Addison Wesley.

Jagoe, A. (2003). *Mobile location services; the definitive guide.* Upper Saddle River: Prentice Hall.

Janssen, M. F. W. H. A., & Wagenaar, R. W. (2003, 1-5 Sept). *From Legacy to Modularity: A Roadmap Towards Modular Architectures Using Web Services Technology*. Paper presented at the Electronic Government, Second International Conference, EGOV 2003, Prague, Czech Republic.

Johansson, N., Lind, T., & Sandblad, B. (2005). Usability in IT systems for mobile work. In J. H. E. Andriessen & M. Vartianen (Eds.), *Mobile virtual work: A new paradigm?* London, UK: Springer.

Johnson, S. P., Menor, L. J., Roth, A. V., & Chase, R. B. (2000). A Critical Evaluation of the New Service Development Process. In J. A. Fitzsimmons & M. J. Fitzsimmons (Eds.), *New service development: Creating memorable experiences* (pp. 1-27). Thousand Oaks: Sage Publications, Inc.

Jones, C., Hesterly, W. S., & Borgatti, S. P. (1997). A General Theory of Network Governance: Exchange Conditions and Social Mechanisms. *Academy of Management Review, 22*(4), 911-945.

Judd, R. C. (1964). The Case for Redefining Services. *Journal of Marketing, 28*(1), 58-59.

Kalakota, R., & Robinson, M. (2002). *M-Business: the race to mobility*. New York: McGraw-Hill.

Kasper, J. D. P., van Helsdingen, P. J. C., & de Vries, W., Jr. (1999). *Service Marketing Management, an International Perspective*. England: John Wiley and Sons Ltd.

Keen, P. W. G., & Mackintosh, R. (2001). *The freedom economy : gaining the M-commerce edge in the era of the wireless Internet* New York: Osborne/McGraw-Hill.

Keeney, R. L. (1999). The Value of Internet Commerce to the Customer. *Management Science, 45*(4), 533-542.

Koeppel, I. (2000). What are Location Services? - From a GIS Perspective. Retrieved 2 February, 2004, from www.jlocationservices.com

Kotler, P. (1997). *Marketing management; analysis, planning, implementation, and control* (9th edition ed.). Upper Saddle River: Prentice Hall.

Lee, J., Siau, K., & Hong, S. (2003). Enterprise integration with ERP and EAI. *Communications of the ACM, 46*(2), 54-60.

Lewis, C., & Norman, D. A. (1986). Designing for Error. In D. A. Norman & S. W. Draper (Eds.), *User centered system design: new perspectives on human-computer interaction* (pp. 411-432). Hillsdale, NJ: Erlbaum.

Liljander, V., Van Riel, A., & Pura, M. (2002). Customer Satisfaction with e-Service: the case of an online recruitment portal. In M. Bruhn & B. Stauss (Eds.), *Yearbook of Services Management 2002 - E-services.*

Linthicum, D. S. (2000). *Enterprise application integration* Reading, Mass: Addison-Wesley.

Maier, M. W., & Rechtin, E. (2002). *The art of Systems architecting.* Boca Raton, FL: CRC-press.

Maitland, C., Van de Kar, E. A. M., Wehn de Montalvo, U., & Bouwman, H. (2005). Mobile information and entertainment services: business models and service networks. *Int. J. Management and Decision Making, 6*(1), 47-64.

Malhotra, A., & Segars, A. H. (2005). Investigating wireless web adoption patterns in the U.S. *Communications of the ACM, 48*(10), 105-110.

Mayer, I. S., & Veeneman, W. W. (2002). *Games in a world of infrastructures : simulation-games for research, learning and intervention.* Delft: Eburon.

Mayhew, D. J. (1999). *The usability engineering lifecycle : a practitioner's handbook for user interface design* San Francisco, CA: Morgan Kaufmann.

Menor, L. J., Tatikonda, M. V., & Sampson, S. E. (2002). New service development: areas for exploitation and exploration. *Journal of Operations Management, 20*, 135-157.

Meredith, J. R., & Mantel, S. J. (2003). *Project Management; a managerial approach.* United States of America: John Wiley.

Mihovska, A., Platbrood, F., Karetsos, G., Kyriazakos, S., Van Muijen, R., Guarneri, R., et al. (2006). Towards the Wireless 2010 Vision: A Technology Roadmap. *Wireless Personal Communications, 42*, 303-336.

Miller, D., & Shamsie, J. (1996). The resource-based view of the firm in two environments: The Hollywood film studios from 1936 to 1965. *Academy of Management Journal, 39*, 519-543.

Mintzberg, H. (1979). *The structuring of organizations: a synthesis of the research.* Englewood Cliffs: Prentice Hall.

Mintzberg, H. (1983). *Power in and around organizations.* Englewood Cliffs: Prentice Hall.

Nah, F. F.-H., Siau, K., & Sheng, H. (2005). The value of Mobile Applications: A Utility Company Study. *Communications of the ACM, 48*(2), 85-90.

Nassimbeni, G. (1998). Network structures and co-ordination mechanisms; A taxonomy. *International Journal of Operations & Production Management, 18*(6), 538-554.

Natsuno, T. (2003). *The i-mode wireless ecosystem.* Chichester: Wiley.

Nielsen, J. (1993). *Usability engineering.* Boston: Academic Press.

Nooteboom, B. (1999). Innovation and inter-firm linkages: new implications for policy. *Research Policy, 28*, 793-805.

Nooteboom, B., & Gilsing, V. A. (2004). *Density And Strength of Ties In Innovation Networks: A Compentence And Governance View.* Rotterdam: ERIM.

OECD. (2007). *OECD Factbook 2007.* London, UK: OECD Publishing.

Orfali, R., Harkey, D., & Edwards, J. (1996). *The Essential Distributed Objects Survival Guide.* New York: John Wiley & Sons.

Papazoglou, M. P., & Georgakopoulos, D. (2003). Service-oriented computing: Introduction. *Communications of the ACM, 46*(10), 24-28.

Parasuraman, A., Berry, L. L., & Zeithaml, V. A. (1991). Understanding Customer Expectations of Service. *Sloan Management Review, 32*(3), 39-48.

Parasuraman, A., Zeithaml, V. A., & Berry, L. L. (1985). A conceptual model of service quality and its implications for future research. *Journal of Marketing, 49*(4), 41-50.

Parasuraman, A., Zeithaml, V. A., & Berry, L. L. (1988). SERVQUAL: a multiple-item scale for measuring consumer perceptions of service quality. *Journal of Retailing, 64*(1), 12-40.

Pedersen, P. E., & Methlie, L. B. (2004, August 22-24). *Exploring the relationship between mobile data services business models and end-user adoption.* Paper presented at the Proceedings of WCC2004-I3E conference, Toulouse, France.

Pfeffer, J., & Salancik, G. R. (1978). *The External Control of Organizations.* New York: Harper & Row, Publisher.

Porter, M. E. (1991). Towards a dynamic theory of strategy. *Strategic Management Journal, 12*(95-117).

Powell, W. W. (1990). Neither Market nor Hierarchy: Network Forms of Organization. *Research in Organizational Behavior, 12*(295-336).

Powell, W. W. (1998). Learning from collaboration: Knowledge and networks in the biotechnology and pharmaceutical industries. . *California Management Review, 40*(3), 228-241.

Powell, W. W., Koput, K. W., & Smith-Doerr, L. (1996). Interorganizational collaboration and the locus of innovation: Networks of learning in biotechnology. *Administrative Science Quarterly, 41*(1), 116-145.

Powell, W. W., & Smith-Doerr, L. (1994). Networks and economic life. In N. J. Smelser & R. Swedberg (Eds.), *The Handbook of Economic Sociology* (pp. 368-402). Princeton: Princeton University Press.

Rogers, E. M. (2003). *Diffusion of innovations* New York Free Press.

Roozenburg, N. F. M., & Eekels, J. (1995). *Product Design: Fundamentals and Methods*. Chichester: Wiley.

Rouse, W. B. (1991). *Design for success: a human-centered approach to designing successful products and systems*. New York: Wiley

Sabat, H. K. (2002). The evolving mobile wireless value chain and market structure. *Telecommunications Policy, 26*, 505-535.

Sage, A. P., & Armstrong, J. E. J. (2000). *Introduction to Systems Engineering*. New York: Wiley.

Salonen, J., & Karjalainen, J. (2003, 23-24 June). *Implementing mobile authentication in electronic transaction services*. Paper presented at the The Second International Conference on Mobile Business, Vienna, Austria.

Schraagen, J. M., Chipman, S. F., & Shute, V. J. (2000). State-of-the-art review of cognitive task analysis techniques. In J. M. Schraagen, S. F. Chipman & V. L. Shalin (Eds.), *Cognitive Task Analysis* (pp. 467-486). Mahwah,New Jersey: Lawrence Erlbaum Associates.

Shannon, R. E. (1975). *Simulation. The art and science*. Englewood Cliffs, NJ: Prentice-Hall.

Shapiro, C., & Varian, H. R. (1999). *Information rules a strategic guide to the network economy* Boston, Mass: Harvard Business School Press.

Shaw, M. J., & Garlan, D. (1996). *Software architecture; perspectives on an emerging discipline* Upper Saddle River: Prentice-Hall.

Shulver, M. (2005). Operational loss and new service design. *International Journal of Service Industry Management, 16*(5), 455-479.

Simons, L. (2006). *Multi-channel services for click and mortars*. Delft, The Netherlands: Delft University of Technology.

Stallings, W. (2002). *Wireless communications and networks*. New Yersey: Prentice Hall.

Stanton, N. A., & Young, M. S. (1999). *A guide to methodology in ergonomics: designing for human use*. London: Taylor & Francis

Stevens, E., & Dimitriadis, S. (2005). Managing the new service development process: towards a systemic model. *European Journal of Marketing, 39*(1/2), 175-198.

Sydow, J., & Windeler, A. (1998). Organizing and Evaluating Interfirm Networks: A Structurationist Perspective on Network Processes and Effectiveness. *Organization Science, 9*(3), 265-284.

Syson, F., & Perks, H. (2004). New service development: a network perspective. *Journal of Services Marketing, 18*(4), 255-266.

Szyperski, C. (1998). *Component Software: Beyond object-oriented programming.* New York, NY: ACM Press / Addison Wesley.

Tafazolli, R., Correia, L. M., & Saarnio, J. (2005, 18 March 2005). Mobile Communications & Technology Platform, Strategic Research Agenda; eMobility Staying ahead! Retrieved 5 October, 2005, from http://www.bit.ac.at/ist/TP/eMobility_SRA_2005_03_18.pdf

Tagg, R., & Freyberg, C. (1997). *Designing Distributive and Cooperative Infromation Systems.* London: Thomson Computer Press.

Tamminen, S., Oulasvirta, A., Toiskallio, K., & Kankainen, A. (2004). Understanding mobile contexts. *Personal and Ubiquitous Computing, 8*(2), 135-143.

Tidd, J., & Hull, F. M. (2003). *Service Innovation* (Vol. 9). London: Imperial College Press.

Timmers, P. (1998). Business Models for Electronic Markets. *International journal for Electronic Markets, 8*(2), 3-8.

Tukker, A., & Tischner, U. (Eds.). (2006). *New Business for Old Europe: Product-Service Development, Competitiveness and Sustainability* Sheffield, UK: Greenleaf Publishing.

Turban, E., King, D., Lee, J., & Viehland, D. (2004). *Electronic commerce 2004: a managerial perspective.* Upper Saddle River: Pearson Prentice Hall.

Uzzi, B. (1996). The Sources and Consequences of Embeddedness for the Economic Performance of Organizations: The Network Effect. *American Sociological Review, 61*, 674-698.

Van de Kar, E. A. M. (2002, 17-19 June). *The Introduction of M-Info - A Teaching Case.* Paper presented at the 15th Bled Electronic Commerce Conference, Bled, Slovenia.

Van de Kar, E. A. M. (2004). *Designing mobile information services; An Approach for Organisations in a Value Network.* Unpublished Doctoral dissertation, Delft University of Technology, Delft, The Netherlands.

Van de Kar, E. A. M., Maitland, C., Wehn de Montalvo, U., & Bouwman, H. (2003, September 30 - October 3). *Design guidelines for Mobile Information and Entertainment Services; Based on the Radio538 ringtunes i-mode service case study*. Paper presented at the Fifth International Conference on Electronic Commerce, Pittsburgh, PA, USA.

Van de Kar, E. A. M., Wang, Y., & Muniafu, S. (2006, 14-16 August). *Mobile Services Used in Unstable Environments: Design Requirements based on Three Case Studies*. Paper presented at the Eighth International Conference on E-Commerce, Canada.

Van den Anker, F. W. G. (2002). *Scenarios@Work: developing and evaluating scenarios related to cooperative work mediated by mobile multimedia communications*. Wageningen, The Netherlands: Ponsen & Looijen BV.

Van Meel, J. W. (1994). *The dynamics of business engineering*. Unpublished Doctoral dissertation, Delft University of Technology, Delft, The Netherlands.

Varshney, U., & Vetter, R. J. (2001, 3-6 Jan). *A Framework for the Emerging Mobile Commerce Applications*. Paper presented at the Proceedings of the 34th Hawaii International Conference on System Sciences, Big Island, Hawaii.

Vassos, T. (1996). *Strategic Internet Marketing*. Indianapolis: Que.

Verbraeck, A., & Dahanayake, A. N. W. (2001). *Building blocks for Effective Telematics Service system Development and Evaluation*. Delft: Delft University of Technology.

Verbraeck, A., Saanen, Y., Stojanovic, Z., Valentin, E., van der Meer, K., Meijer, A., et al. (2001). What are building blocks? In A. Verbraeck & A. N. W. Dahanayake (Eds.), *Building blocks for Effective Telematics Service system Development and Evaluation*. Delft: Delft University of Technology.

Verbraeck, A., Tewoldeberhan, T., & Janssen, M. F. W. H. A. (2006). E-Supply Chain Orchestration. *Encyclopedia of E-Commerce, E-Government and Mobile Commerce*.

Vermeulen, P., & van der Aa, W. (2003). Organizing Innovation in Services. In J. Tidd & F. M. Hull (Eds.), *Service Innovation* (Vol. 9, pp. 35-53). London: Imperial College Press.

Wallnau, K. C., Hissam, S. A., & Seacord, R. C. (2002). *Building Systems from Commercial Components* Boston, MA: Addison-Wesley.

Wang, Y., Van de Kar, E. A. M., Meijer, G., & Hunteler, M. (2005, 11-13 July). *Improving business processes with mobile workforce solutions*. Paper presented at the Proceedings of the International Conference on Mobile Business, Sydney, Australia.

Wehn de Montalvo, U., Van de Kar, E. A. M., & Maitland, C. (2005). Resource-based interdependencies in value networks for mobile e-services. *International Journal of E-Business Research, 5*(1), 1-20.

Weill, P., & Vitale, M. R. (2001). *Place to Space; Migrating to eBusiness models*. Boston, MA: Harvard Business School Press.

Weinberg, G. M. (1975). *An Introduction to General Systems Thinking*. New York: Wiley.

Wernerfelt, B. (1984). A resource-based view of the firm. *Strategic Management Journal, 5*, 171-180.

Wiklund, M. E. (1994). *Usability in practice; gow companies develop user-friendly products*. Boston: AP Professional.

Williamson, O. E. (1985). *Economic institutions of capitalism; firms, markets, relational contracting*. New York Free Press.

Winston, W. L. (2004). *Operations Research: Applications and Algorithms* (4th ed.). Belmont, CA: Brooks/Cole - Thomson Learning.

Yuan, Y., & Zjeng, W. (2005). *From Stationary Work Support to Mobile Work Suppport: A Theoretical Framework*. Paper presented at the the International Conference on Mobile Business, Sydney, Australia.

Zeigler, B. P., Praehofer, H., & Kim, T. G. (2000). *Theory of Modeling and Simulation. Integrating discrete event and continuous complex dynamic systems* (2nd ed.). San Diego, CA: Academic Press.

Zeithaml, V. A., Bitner, M., & Gremler, D. D. (2006). *Services Marketing; Integrating Customer Focus Across the Firm* (4th ed.). New York, NY: McGraw-Hill.

Zigurs, I., & Buckland, B. K. (1998). A theory of task/technology fit and group support systems effectiveness. *MIS Quarterly, 22*(3), 313-334.

Contributors

The persons who contributed to this book are highlighted below.

Chapter authors

Chapter 1, Introduction
Els van de Kar and Alexander Verbraeck

Chapter 2, Service Concept
Els van de Kar and Mirjam Huis in 't Veld

Chapter 3, Technical Architecture
Marijn Janssen and Alexander Verbraeck

Chapter 4, Organizational Architecture
Els van de Kar and Carleen Maitland

Chapter 5, Design Approaches for Service Systems
Els van de Kar, Alexander Verbraeck, Mirjam Huis in 't Veld, Wieke Bockstael-Blok and Yan Wang

Chapter 6, Mobile Information Service System Design Approach
Els van de Kar

Chapter 7, MIES on the Campus
Els van de Kar

Appendix A, Unified Modelling Language
Marijn Janssen

Staff contributors

Dr. ir. Wieke Bockstael-Blok presently works as consultant at TNO in the Business Unit Work and Employment. Until 2006 she worked as Assistant Professor at the Delft University of Technology, were she also completed her PhD research in 2001, titled Chains and Networks in Multimodal Passenger Transport: Exploring a design approach. Her research interest lies in improvement of the performance of organizations and networks of organizations from a systems' perspective in which 'soft' theories on and approaches to organizational change and 'hard' quantitative systems analytical tools and techniques such as process modelling and simulation, are combined.

Mirjam Huis in 't Veld, MA, currently works as a researcher at TNO in the Business Unit Human Factors. Until 2005 she worked as an assistant researcher at the Delft University of Technology, Faculty of Technology, Policy and Management. Her research focuses on the ICT support of groups in distributed settings. In her work she combines theories within the area of social psychology, organisation behaviour, Computer Supported Co-operative Work (CSCW) and ICT design and evaluation.

Dr. Marijn Janssen is an Associate Professor and director of education of the SEPAM (Systems Engineering, Policy Analysise and Management) Master at Delft University of Technology, Faculty of Technology, Policy and Management. His research is focused on ICT-service engineering of public service networks (e-government). His research is published in a large number of refereed conference proceedings, book chapters, and international journals.

Dr. Els van de Kar is Assistant Professor in the Faculty of Technology, Policy and Management (TPM) at Delft University of Technology in The Netherlands. She received a PhD. in systems engineering for her research on a design approach for mobile information services in which the customer, technology and organizational issues are balanced. Van de Kar has more than 10 years working experience in industry at the Dutch telecommunication operator KPN Telecom and as independent consultant. Van de Kar's research focuses on service design, and in particular the design of mobile service systems.

Dr. Carleen Maitland is Assistant Professor in the School of Information Sciences and Technology at The Pennsylvania State University in the USA. She received a doctorate in 2001 from Delft University of Technology in the Netherlands, where she was an Assistant Professor prior to joining Penn State. Her research interests include the effects of new mobile and wireless technologies on organizations and industry structure, as well as the role of national policies in bringing about these effects. She has published in journals such as Telecommunications Policy, Information Economics and Policy and Communications and Strategies.

Prof. Dr. Ir. Alexander Verbraeck is chair of the Systems Engineering Group of the Faculty of Technology, Policy and Management of Delft University of Technology, and a part-time full professor in supply chain management at the R.H. Smith School of Business of the University of Maryland. He is a specialist in discrete event simulation for real-time control of complex transportation systems and for modelling business systems. His current research focus is on development of generic libraries of object oriented simulation building blocks in C++ and Java, which can be used to support designing, testing and training complex service systems.

Dr. Yan Wang was a researcher in the Systems Engineering group at the Faulty of Technology, Policy and Management (TPM) of Delft University of Technology, the Netherlands during the writing of this book. He completed a PhD in 2008 on mobile business, aimed at designing adaptive business processes using a simulation approach. Currently, Yan Wang is employed by Accenture.

Student contributors

The "Guido the Guide" design example was developed by René Putman and Ron Sperling as the final assignment in the course spm4140 – Service Systems Engineering in 2003-2004. Section 1.6 is based on the text of their final assignment.

The design example "Port Emergency Harbour System" was developed by Nitesh Barosha and Laurens Waling as the final assignment in the course spm4140 – Service Systems Engineering in 2004-2005. Section 1.7 is based on the text of their final assignment.

The "Buddy on Demand" design example was developed by Mark de Reuver and John van der Sar as the final assignment in the course spm4140 – Service Systems Engineering in 2003-2004. Section 2.8 is based on the text of their final assignment.

The "*SWIFT* Band" design example was developed by Freek Faber and Wouter van Lelyveld as the final assignment in the course spm4140 – Service Systems Engineering in 2005-2006. Section 2.9 is based on the text of their final assignment.

The "Finding disaster victims service" design example was developed by Tim Hehenkamp and Erik Zwaneveld as the final assignment in the course spm4140 – Service Systems Engineering in 2004-2005. Section 3.9 is based on the text of their final assignment.

The "Last minute business travel changes" design example was developed by Gabriel Marcuzzo do Canto Cavalheiro and Krupa Iyer as the final assignment in the course spm4140 – Service Systems Engineering in 2003-2004. Section 4.6 is based on the text of their final assignment.

The "people tracing service" design example was developed by Jessica Groeneweg and Jasper van Waalwijk as the final assignment in the course spm4140 – Service Systems Engineering in 2004-2005. Section 4.7 is based on the text of their final assignment.

The "Cycling tours in Europe" design case was developed by Kanwal Khan and Kim Putter as the final assignment in the course spm4140 – Service Systems Engineering in 2003-2004. The case in the appendix is based on the text of their final assignment.